CREATIONISM'S UPSIDE-DOWN PYRAMID

How Science Refutes Fundamentalism

LEE TIFFIN

 Prometheus Books

59 John Glenn Drive
Amherst, New York 14228-2197

Published 1994 by Prometheus Books

98 97 96 95 94   5 4 3 2 1

Library of Congress Cataloging-in-Publication Data

Tiffin, Lee, 1919–
    Creationism's upside-down pyramid : how science refutes fundamentalism / Lee Tiffin.
          p.   cm.
    Includes bibliographical references and index.
    ISBN 0-87975-898-8 (alk. paper)
    1. Creationism—Controversial literature. 2. Religion and science—United States. 3. Religion and politics—United States. I. Title.
BS651.T54   1994
231.7'65—dc20   94-15920
                                                                CIP

Printed in the United States of America on acid-free paper.

Contents

Part Two: Contrasts and Challenges: Science versus Creationism

Part Three: Public Concerns and Responsibility

Acknowledgments

The research and writing for this book have extended over most of a decade. I owe special thanks to my brother, Richard Tiffin, for the photography and glossy prints of illustrations used in the book.

To many individuals I must express thanks for prompt responses to requests for technical information, bibliographic references, reprints of articles, manuscript review, and editing. Among the most helpful of the scientists and other professionals I contacted are the following: Alan Taylor, Nancy Wells, Patricia Tiffin, Eugenie Scott, Robert Fraser, Rufus Chaney, Rod Norum, Don Latham, Richard Anthes, Christopher Toumey, and Edd Doerr.

All the assistance given by my wife, Donna, cannot be measured. Without her willing embrace of increased household responsibilities, this book likely would not exist. And to our children, Sharon, Tim, and Sue (his wife), I express gratitude for their help and understanding. To them and their children I dedicate this book, with hopes for their continuing intellectual growth as true friends of science and benefactors of society.

Lee Tiffin
June 1993

Introduction

How much more exciting to be hit on the head by a fantastic piece of sky than by a mere acorn of truth from a local tree.

If the fable of Chicken Little were altered to reflect certain fundamentalist crusades, the dedicated chick would be in a lifelong battle with Mister Owl, the skeptical investigator who traced a mysterious "falling sky" to a single acorn that fell from a tree. Moreover, the crusading chicken, like today's leading creationists, would conduct seminars and debates, write books and classroom aids, establish institutes and conduct so-called field and library research—all for one purpose: to prove that it really was a piece of sky that fell.

And the disciples of the crusading chicken would preach everywhere that the falling sky was a divine act. Moreover, its collapse was not merely a local phenomenon, but a universal event that raised floodwater above all the highest mountains.

Conservative Christian leaders are variously called "creationist," "far-right," "young-earth, and "flood-geology" fundamentalists. They and their followers are determined to carry Bible-based teachings everywhere. They retell and amplify the biblical stories of the first eleven chapters of Genesis. The narratives of creation and the Genesis flood are quite brief, but creationist versions include elaborate detail of a literal six-day creation and a vaporous ocean in the sky that collapsed to drown animals and that caused profound geological changes on earth. Creationists give their

11

fictions pretentious names such as "creation science" and "scientific creationism."

Increasingly, scientists and educators are concerned because falling-sky fundamentalists are grasping for greater political power and insisting that they should present their views in tax-supported institutions. Basically, these crusaders assert that their stories represent true science and deserve to be taught in public school science classes. A consensus among many in the scientific and educational communities and in mainline church bodies is that so-called "scientific creationism" based on Bible stories does not deserve a place in the public school science curriculum.

In recent years, creationists have invoked academic freedom, balanced treatment, and equal time in efforts to influence public education policy. They argue that to ensure balance and fairness they should have opportunity in public schools to present their views. Moreover, they declare, nothing could be fairer than letting people hear both sides of issues before making their own decisions.

That challenge raises a dilemma for creationists. If people have the right to hear the fundamentalist message, how much of it are creationists willing to tell? Do they insist that people hear and accept doctrines by faith, but not critically examine what they must believe? Often creationism's most impressive statements are like masks: they catch attention and arouse curiosity—but never reveal what is below the surface.

My goal in this book is to lift and look under creationists' masks. Part One provides a guide to better understanding of what they believe and what strategies they employ to achieve their sectarian religious goals. I discuss basic creationist assumptions, their un-scientific methods, and their remarkable ability to manipulate virtually everything: stars, starlight, atmosphere, mountains, oceans, rivers, plants, animals, and people. I clearly show why creationists should not be allowed to teach their "science" and why their curriculum guides and religious materials should not be adopted for use in the public schools. I emphasize, however, that nowhere have I denied anyone's rights to personal and private convictions, nor has the legitimate practice of religious rites been questioned.

Part Two explains the scientific information introduced in Part One to provide a foundation of physical measurements and numerical data from recognized standard and classical authorities. Readers willing to think through the few moderately technical areas in Part Two will be able to confirm that "creation science" rests on surmise and guessing, not on solid, verifiable evidence. They will understand how leading creationists have discredited themselves as scientists and have no remedies for America's illiteracy in science.

Part Three shifts to politics and religion, explaining why we should be concerned not only about the intellectual crisis creationism poses for society but also for the climate of sanctimony, where sectarian groups fashion religious tests for office seekers and politicians commit themselves to the doctrinal triumph of sectarian religious opinion.

I wrote this book hoping that it may in some way counter the influence of religious zealots and at the same time stimulate appreciation for genuine science. My greatest concern has been for public school physical and life science courses and related programs. I hope to contribute to the following goals:

1. The public will gain a clear understanding of the pretensions of "creation science" and its so-called scientific goals for society.

2. More people will understand that if what "creation science" teaches about the distribution and role of water, light, and other forces on the primitive earth had been true, life as it is known today could not exist.

3. More scientists will understand the ever-increasing ways creationists find to obscure and reject clear facts of science, will consider the mounting illiteracy in science caused by such rejections, and will join the forces of resistance.

4. Contradictions of earth history and natural laws by creationists will be more widely recognized by teachers, parents, principals, school boards, and librarians to prevent suppressive policies and unfortunate selections of textual materials that distort and dilute science and related programs.

5. Public servants will embrace separation of church and state and, despite political pressures and even religious persecution, be ready to resist every deliberate attempt to subvert that separation.

6. School administrators faced with difficult dilemmas of church and state will find needed answers and motivation for valid decisions that do not violate the public trust (e.g., declining offers of sectarian religious materials and lecturing evangelists who preach their "true" facts of science).

In their writing and debating strategy, creationists eagerly challenge opponents on details of genetics and mutation, new species and body forms, and fossils. This book does not respond directly to those challenges. Why, I ask, should anyone take seriously the creationists' arguments about living things if their earth science models would have made life impossible?

For instance, creationists are quick to decide that eyes could not have evolved. I do not take up that debate. Several issues are more crucial than evolution of eyes. For example, a basic creationist belief is that an incredibly large quantity of water vapor filled their preflood sky. The physical and optical density of their massive water shell would have blocked incoming light throughout the preflood period. I ask, what use would animals have for eyes? Debate about eyes on a dark planet unable to support life is futile.

This book focuses on common elements necessary for life. The sun provides light and heat; the earth supplies water, gases, and minerals. These factors existed before living things. A derived product, food, also must be included as the sixth essential for sustaining life.

In manipulating these six essentials, leading creationists have published serious errors. I attempt to do what they are not willing to do—analyze objectively their published models and identify conceptual errors.

The central question of later chapters is whether any organism could have arisen and survived under conditions of the creationist model. I intend to show that where creationist imaginations have ranged far from principles on which life depends, they have created formulas under which life could not exist. This information will help readers answer two pertinent questions: Do creationists fully embrace the principles and methods of science? Should creationists and the curricular programs they offer be given a place in public schools?

If the fable of Chicken Little is a parable of human folly, it also serves as a model of learning and of wisdom. The chicken and friends made an exemplary choice: they accepted the truth of the acorn. Thus, they could dismiss the fanciful story and their once-urgent mission to "run and tell!" We presume that finally, in barnyard and field, they returned to pursuits natural and essential.

Part One

Issues and Challenges

1

Creationist Assumptions and Methods

Analogy and Overview

The creationist world view assumes that there once was a global flood and that extraordinary climatic conditions preceded it. Creationists believe that the biblical flood is a pivotal event in history and that a complete understanding of it could resolve many controversies that now exist between themselves and their opponents. Their list of opponents, remarkably long, includes biologists, geologists, evolutionists, uniformitarians, atheists, materialists, naturalists, secular humanists, and satanists.

The primary assumptions creationists have made concerning a global flood and preflood era may be visualized as building stones in a pyramid. The pyramid does not rest on a broad base of many disciplines. Nor does it ever rise to a capstone of coherent theory or scientific law.

Rather, the builders place a single postulate at the base of an inverted pyramid. Upon that base they add other assumptions, and upon those they add still other layers. Ultimately, they believe, these constitute overwhelming scientific knowledge. Scientists reject the first postulate: the belief that sacred writings describe earth, biological, and stellar origins as having occurred within the past 10,000 years. They also reject creationism's rickety structure, which is, in fact, a pyramid of surmise and guessing, not scientific knowledge.

The inverted pyramid (see fig. 1) symbolizes confusion. The

endless gathering of stones and the need for props of rhetoric represent creationism's frenzied, opportunistic activities. The labels on stones indicate subjects to be discussed in coming chapters.

Although creationists embrace an upside-down view of nature, they sometimes recognize a building block that misrepresents facts. What should they do when biblical writers (or they, as "scientific" investigators) have made serious mistakes?

A sensible course of action would be to reject a stone outright and be rid of it. Creationists, however, would rather make revisions and renovations to keep a questionable stone in place. The offending stone is relabeled, and additional props are arranged to give imagined stability. Rhetoric apparently is unlimited; but if all the props creationists imagine were set in place, much of the pyramid's surface would be obscured.

In one example of a major biblical revision, two leading creationists decided that the food-for-survival plan for animals on Noah's ark (Genesis 6:21, 22) was not feasible. They offered their own interpretation: the animals hibernated. I discuss this topic further in chapter 11.

Another problem is the Paluxy mantracks assumption. Creationists finally realized—after receiving valuable assistance from scientists—that purported human footprints in the Paluxy riverbed in central Texas were not human. They were tridactyl footprints altered by natural causes, chiefly erosion and infilling of dinosaur tracks by water action.

Creationists have reacted to the Paluxy fiasco not only with half-hearted retractions and withdrawals of literature and films, but also with reassurance for the faithful by promising continued research on the Paluxy River. They displayed neither obvious good will nor cooperation with scientists who gave rational reviews of the Paluxy data.

By crossing out the word "mantracks" and relabeling, creationists have created a new pyramid block: Paluxy Mystery. Their mystery has kept the mantrack question open and encouraged believers. The Paluxy Mystery also has provided grist for inveterate creationist writers. I discuss the Paluxy issue in chapter 10.

Fig. 1. Creationism's Upside-Down Pyramid

Crucial Creationist Doctrine: Floodwater Sources

I chose to examine critically the Genesis flood because creationists consider this event to be a dominant force in shaping earth's geology and biology. In developing the story, they base much of their exposition on Bible verses that, they believe, identify flood-water sources. One statement (Genesis 1:7) refers to an uplifted water shell (creationism's fictitious vapor canopy). The other verse (Genesis 7:11), as interpreted by creationists, identifies a source confined in underground reservoirs until released in the flood.

Taking biblical statements as literal truth, creationists have proceeded to develop "flood science." However, scientifically proving the existence of the flood involves formidable challenges. Not only must the floodwater sources be verified, but scientific mechanisms for their release must be identified. Creationists fulfill neither requirement. Nevertheless, they postulate canopy and cavern sources—to be accepted on faith—then attempt to support the mysterious water release and worldwide flooding from current knowledge in the earth and biological sciences.

What are the fundamental beliefs that underlie creationism's "flood science"? I present the following list to describe tenets of deluge doctrine. Most of the tenets are based on biblical writings, as interpreted by Whitcomb and Morris[1]:

1. Vapor canopy collapse and forty-day rainfall

2. Explosive releases of heated underground liquids and gases

3. Floodwater higher than all mountains

4. Stripping of vegetation and soils, with subsequent transport and submergence by water

5. Massive water erosion of transportable subsoil and rocks with consequent deep burial of organic materials, accounting for earth's inventory of fossil fuels—all occurring in the "flood year"

6. Human and animal drownings followed by water distribution of their bodies, accounting for all such fossils in the geologic column

7. Massive freezing of Arctic animals at some flood stage

8. An ark-rescue of selected humans and animals, the antecedents of current populations

9. Earth's post-flood human inhabitants descend exclusively from Noah's sons

10. Subsequent biblical family lines and related sacred history

Creationists recognize that floodwater sources are indispensable elements in their flood story. If those sources did *not* exist, then creationists must forfeit their most basic assumptions of flood-related geology and biology.

Creationism's Cardinal Admissions

Creationists present themselves as scientists and propose the phenomena cited above as basic tenets of their "science." But what scientific evidence verifies the existence of the canopy?

The foremost leaders in the creationist movement answer the question with this apology:

> Although we can as yet point to no definite scientific verification of this pristine vapor protective envelope around the earth, neither does there appear to be any inherent physical difficulty in the hypothesis of its existence, and it does suffice to explain a broad spectrum of phenomena both geological and Scriptural.[2]

Scientists readily agree that creationists have no scientific evidence for their vapor canopy. Regardless, creationists base a long and unconvincing story on the surmise. In addition, creationists expect people to believe that they are scientists and to accept unquestioningly their story as truth.

Scientists may consider it a fact that creationists never perceived intellectual difficulties in the surmised existence of their canopy. Their lack of perception could indicate lack of scholarship.

Despite the absence of confirming evidence, the canopy surmise proposed by creationists satisfies *their* needs. Implications of the statement "and it does suffice to explain . . ." are indeed remarkable. This means, by definition (*Webster's New Universal Unabridged Dictionary* for "suffice"), that the canopy conjecture can "answer the purpose or requirements of" creationists in explaining selected phenomena. The canopy surmise may satisfy creationist standards, but it cannot measure up to the criteria and standards of science.

Preview: Creationist Assumptions

Creationists strongly desire to be recognized as scientists. They present themselves, however, as creation scientists who possess deep insight and unique knowledge about the true foundations of science.

What is that special knowledge? To answer that question, I summarized from their literature seven propositions. The last sentences of proposition 7 refer to alleged ark experiences taking place in a flood year. All other statements refer to biblical preflood times. The basic assumptions are as follows.

1. The Bible expresses historical truth about a massive water shell lifted above earth at creation (Genesis 1:7). The elevated water was always in gaseous form until it collapsed in a universal flood.

2. The biblical statement "fountains of the great deep burst forth . . ." (Genesis 7:11, Revised Standard Version [RSV]) refers to underground reservoirs of hot, pressurized liquids released in the Genesis flood or the uplift of ocean floors.

3. The vapor canopy rested upon a supporting layer of heavier gases for several thousand years before it collapsed in the flood.

4. The vapor canopy captured all ultraviolet and infrared radiation but allowed penetration of visible wavelengths.

5. Energy intercepted and transmitted by the vapor canopy warmed the whole earth and virtually eliminated seasons, even at the poles.

6. Significant atmospheric movements (cyclonic fronts, wind-storms) were absent throughout the preflood period.

7. Provisions of every sort were stored on an ark to serve as food for a year-long survival of selected humans and animals.

Some creationists, doubting the feasibility of the food-for-survival plan, offer an alternative. Hibernation would eliminate the need to feed the animals during the flood year.

Since no definitive evidence can be offered, believers must accept these assumptions by sheer acts of faith. Not mentioned in the above list are many alleged events in the so-called flood year and postflood period.

Creationism's Unscientific Methods

The foregoing propositions raise serious questions about creationism and its scientific credibility. As emphasized above, faith is the most basic requirement for a creationist. Apparently, for many the imagined virtue of their faith is in believing what cannot be verified.

Creationist errors in methodology raise other serious problems. Their simple standard is that no assumption requires scientific support before the next one is offered. For example, creationists believe that their elevated water shell was literal historical fact. Before that assumption can be verified they declare that the elevated water was exclusively in vapor form. The vapor shell is then imagined to have been layered above other atmospheric gases according to molecular weight. However, a strict weight distribution would have brought death by carbon dioxide asphyxiation of animals and humans. Creationists then assume that their layered atmosphere remained *stationary* and its gases *unmixed* for several millennia; thus, there could not have been any winds or diffusion of gases.

Scientists reject creationist manipulations of matter. Blocking the movement of molecules and the movement of atmosphere denies the operation of basic laws of thermodynamics.

Creationism's assumption pyramid has serious flaws; first guesses do not confirm later guesses, nor do later ones substantiate the former. Depending as they do on surmise and guessing, and at the same time contradicting physical laws, creationists cannot expect to achieve scientific credibility.

Notes

1. J. C. Whitcomb and H. M. Morris, *The Genesis Flood* (Phillipsburg, N.J.: Presbyterian and Reformed Publishing Co., 1961).

2. Whitcomb and Morris, 1961, p. 241. Creationist views that I have examined were found mainly in six books: Whitcomb and Morris, 1961; J. C. Whitcomb, *The World That Perished* (Grand Rapids, Mich.: Baker Book House, 1973); H. M. Morris (ed.), *Scientific Creationism,* General edition (El Cajon, Calif.: Master Books, 1974); H. M. Morris, *The Beginning of the World* (Denver Colorado: Accent Books, 1977); J. D. Morris, *Tracking Those Incredible Dinosaurs and the People Who Knew Them* (San Diego, Calif.: CLP Publishers, 1980); and J. C. Whitcomb, *The Early Earth* (Grand Rapids, Mich.: Baker Book House, 1986).

2

Creationism's Earth and Stellar History

Spurious Age: Things Old Are Young

Apparent age is a vital stone in the creationists' inverted pyramid. "Appearance of age" is a convenient explanation for such things as planets and stars that are billions of years old, but in the creationist view have existed less than 10,000 years.

Many writers have used the term *omphalos* (the Greek word for "navel") to describe creationist arguments about apparent age.[1] The source of the metaphor is a nineteenth-century theological treatise[2] that stated that God created Adam complete with a navel, indicating attachment to a mother. Creationists concluded that God had created Adam, as well as earthly and stellar objects, with an appearance, or evidence, of maturity and age. The Gosse hypothesis is a first-tier stone in creationism's upside-down pyramid.

Most modern fundamentalist writers are aware of Gosse's *omphalos* argument but are not willing to embrace the specific navel hypothesis. Whitcomb declared[3] that nothing evolved, but that everything was created full-grown and functional. An editor of the *Bible-Science Newsletter*[4] offered several examples of apparent age. He suggested that "sea water may have had elements dissolved in it. . . ." Also, because "stars were created for the purpose of being seen on Earth," stars and the starlight radiating from them to earth were created at the same time. It is strange that billions of stars allegedly created to be seen cannot be viewed without the aid of telescopes.

Creationist Time Scale and Physical History

Many creationists believe in a literal six-day creation of the universe 6,000 to 10,000 years ago: ". . . Creation occurred thousands (but not tens of thousands) of years before 4004 B.C."[5]

Young[6] listed thirty-seven authorities, whose dates for creation range from 6,984 to 3,616 B.C. Ussher's date, 4004 B.C., was commonly given in the King James Bible until early in this century.

Creationists divide earth's history into three periods (see fig. 2). Period I, from creation to Noah's flood, lasted a minimum of 1,656 years according to strict biblical chronology (Genesis 5:3–28; 7:6). Period II covered 371 days as follows: IIa, 40 days; IIb, 110 days; IIc, 221 days.[7] Period III has endured more than 4,300 years, if biblical statements in Genesis 7:6 and 8:13 and the estimate for Noah's birth in 2948 B.C.[8] express historical facts.

Scientific articles on specific isotope quantities, their decay rates, the quantities of daughter elements, and other data allow estimates of the age of elements and thus the inferred age of the universe.[9] The most comprehensive source concerned with ages of the earth and stars is the recently published book by Dalrymple.[10] Radiological and chemical examinations of materials on earth give an age of about 4.7 billion years. Measured radiations from stars provide an estimate of more than 10 billion years for the universe.

Earth's Water Inventory and Distribution

Earth scientists can reliably estimate the volume of water on the planet. The U.S. Geological Survey[11] estimated that earth's water in the oceans, ice fields, rivers, lakes, rocks, soils, and atmosphere equals 326 million cubic miles (1.36×10^9 km^3). The quantity in the oceans (97 percent of the total) is 316 million cubic miles.

The ocean's volume varies as ice fields build and recede through ice age cycles. Even with such changes, the distribution of water, and equilibria between its phases, has probably existed through long periods in the system known today, wherein great storms generate rain and icy forms of precipitation globally. Such activity has not been confined to the last 10,000 years.

Although creationists apparently accept the present estimated

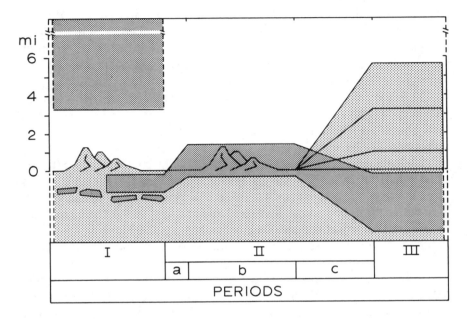

Fig. 2. Diagrammatic representation of the creationist model of earth's physical history. Broken lines indicate unspecified depths of land, sea, and vapor canopy. Periods (not to scale): I. Preflood, allegedly a geologically quiet period with reservoirs of pressurized liquids and gases under land and shallow seas, a universally warm climate, low ice-free mountains, and a massive water vapor shell. II. Flood Year, a cataclysmic interval involving (a) collapse of vapor shell, uplift of ocean floors, ocean overflow, and universal flooding; (b) floodwater sustained above the highest mountains; (c) collapse of ocean floors, continental drainage into deepened ocean basins, mountains elevated to present heights. III. Postflood, a noncataclysmic period extending to the present. The highest horizontal line in period III represents the elevation of Mt. Everest. See text for time-frame estimates by creationists.

quantities of earth's water, their views on its distribution are radically different. Not only do they reject earth history beyond about 10,000 years, they impose on their model strict constraints that must hold throughout Period I. Creationists believe that, at the beginning, the Deity divided earth's water, lifting part above the firmament (an undefined depth of sky) and leaving the remainder in shallow seas and underground reservoirs. These three sources allegedly were reserved in place and brought together in a universal deluge. I discuss the surmised locations of the atmospheric water shell, its mass, and form in later sections.

The biblical narrative (Genesis 7:19) relates that floodwater covered "all the high mountains under the whole heaven." But how much additional water was needed? Computation, using the sphere-volume formula, $4/3\ \pi r^3$, where earth's radius is 3,959 miles (6,371 km) and Mt. Everest's height is 29,029 feet or 5.5 miles (8,848 m), shows that an additional 1,085 million cubic miles would have been required to raise sea level above Mt. Everest. The ratio 1,085:316 gives a water volume that is 3.4 times greater than earth's available ocean water inventory. Obviously, creationists must find additional water—or else drastically manipulate earth's topography.

A similar computation of water needed to raise sea level above Mt. Ararat, where Noah's ark is said to have landed, gives 2.2 additional ocean volumes. The present height of mountains is not the idea of "high mountains" that some leading creationists have in mind. One of the central stones in creationism's inverted pyramid (see fig. 1) represents the alleged floodwater reaching above *low* mountains.

Water Vapor Canopy: Assumptions About Quantity

The flood story involves an immense atmospheric water canopy elevated at the time of creation. What compelling scientific evidence confirms its existence? Creationists can offer none. The Bible, taken literally, speaks of a visible, not molecular, form of water.

Creationists disclaim any knowledge of a mechanism for the alleged separation and elevation of canopy water at creation: "Whether terrestrial heat was instrumental or extraterrestrial forces of some kind or whether solely due to creative fiat, we do not know."[12]

Despite their confessed inability to identify the mechanism for canopy water elevation, or even to verify the canopy's existence, the creationists continue to develop the very complex story. The surmised canopy is at present a basic tenet of fundamentalist doctrine where it assumes a central role both in alleged preflood life and in later flood geology.

If a canopy existed, how much water did it hold? Whitcomb and Morris assumed that at temperatures above water's boiling point "a tremendous amount of water vapor" could be sustained "above the stratosphere."[13]

Other statements further emphasized the great quantity of precipitable water in the canopy. "With much of the earth's water above the atmosphere, the oceans were much smaller, probably occupying only about half the earth's surface."[14] These authors further stated that ". . . much of the waters of our present oceans entered the oceans at the time of the Flood."[15]

Walter Lammerts gave a more definite estimate of canopy water in a discussion of the effects of ocean salt on plant survival: "In fact, on the basis of the canopy theory, we would most certainly expect that the salt content of the ocean before the flood would be diluted, perhaps by one-half."[16] Lammerts' idea seems clear— mixing equal parts of canopy and ocean water would dilute the ocean to half its preflood salt concentration. Whitcomb and Morris obviously approved. They indexed "canopy" with this information: "caused change in oceanic salinity."

Whitcomb and Morris[17] also assumed a "sudden increase of 30 per cent" in the alleged preflood ocean. Dillow[18] proposed that forty feet of precipitable water could be held above the atmosphere.

Creationists have other problems besides holding massive quantities of water in the sky. They cannot explain how plants— even salt-tolerant species—could survive deep submergence in diluted ocean water. Nor have they shown how any land plant could survive darkness or the pressure for most of a year.[19]

Hollow Chamber at Earth's Core

Creationists may require only a minimal amount of information for the creation of religious doctrine. A simple adverb tells the location of heaven. What Sunday School child has not learned that heaven is up? Such a location seems intuitive for believers, particularly for those pointing upward from what seems to be a flat earth. Do fundamentalists ever teach their children that up, from the opposite side of the world, could locate heaven in a very different direction?

Perhaps those who take Bible words literally may now understand that the adverb "down" locates another mysterious place exactly at the center of the earth. We are indebted to Osmon for disclosing a remarkable aspect of Henry Morris's geology. After a lecture on October 24, 1986, Henry Morris responded to a question about the "bottomless pit" of Revelation 9:

> How can there be a bottomless pit? Well, very simple. Always, whenever Hades or Sheol is referred to in the Bible, it's always down in the earth, the depths of the earth. So right there in the center of the earth, apparently, there's a great opening that we can't really deal with in terms of our seismic instruments or other instrumentation. But it apparently is there. You can take the Bible to mean what it says, and I think we have to do that. And of course, since it is at the center of the earth, every boundary has a ceiling. It doesn't have a bottom; it's a bottomless pit.[20]

Henry Morris affirms that his answer is "very simple." Is it likely that his answer will be taken as a new doctrine in "creation science"? He obviously believes that he has identified the location of a great cavern in earth's core, which is beyond scientific testing. One might wonder how satisfied Morris may be with his "geological" data on the exact site of hell.

The statement that "the great opening" can't be dealt with by modern technology seems designed to build confidence in believers, assuring them that the doctrine is biblically sound and beyond contradiction. Although Christendom has been denied this information for nearly 2,000 years, I suspect that few seismologists will be intimidated.

In contrast to Morris's pronouncement, current theoretical studies and seismology show that earth's outer core is fluid and interacts with the lower mantle, but the central core is solid.[21] From high-pressure melting experiments performed on iron, the best temperature estimate for earth's central core is $6,900° \pm 1,000°$ K.

Thoughts of extremely high temperatures and pressures at the center of the earth will undoubtedly stimulate the imaginations of creationists. Some followers may conclude that Henry Morris was divinely inspired to identify the actual site of hell. One may seriously wonder what this addition to the store of "creation science" will mean for believers. Will they find some morbid satisfaction in this new revelation that tells precisely where their enemies are to burn forever? One might question also whether this might be a part of creationist earth-science programs if their curriculum should become established in the public schools.

Mars, Moon Craters, and Warring Spirits

Henry Morris disclaimed belief in ancient myths and pagan astrology by writers such as Immanuel Velikovsky. "Certainly the physical stars as such can have no effect on the earth, but the evil spirits connected with them," he informs us, "are not so limited."[22]

Morris appears reluctant to relinquish, by default, that marvelous territory of stars and asteroids and the wars of angels and devils to the secular and pagan mythmakers. Such sacred, biblical territory belongs to creationists. Some Bible passages, he guesses, may be "simply figurative, but then again they may not."[23] Thus Morris leaves open the possibility of catastrophes among stars and planets taking place after the biblical creation.

Angels allegedly deserve further research as the cause for disturbances in stars and planets.[24] Morris assumed, of course, that stars and planets were created complete and perfect on the fourth day of creation. In his view, "Angels, both good and bad, can be shown Biblically to have considerable knowledge and power over natural processes and, thus can in many cases either cause or prevent physical catastrophes on earth and in the heavens."

In reference to "heavenly warfare," Morris stated that asteroids and planets in the solar system ". . . would be particularly likely to be involved, in view of the heavy concentration of angels, both good and evil, around the planet Earth."[25] Have creationists ever produced evidence for a "heavy concentration of angels" around earth?

In his own research into disturbances, Morris suggested several catastrophes, beginning with ". . . fractures and scars on the moon and Mars."[26] Apparently he rejects all natural, dynamic processes such as volcanic eruptions or incoming matter that causes impact craters. Rather, evil powers in the spirit world apparently are hurling matter about the universe to disorganize and deface the alleged perfect creation.

Morris further mentioned "well documented. . . . 'U.F.O.' sightings."[27] These were associated with "occultic influences and tendencies" because rulers of darkness, he believed, "are increasingly imaginative" in battling for men's minds.

Other assumed disturbances fall into the same class. Morris mentioned "shattered remnants" of a former planet "that became the asteroids."[28] He didn't name the asteroids. Morris also called attention to "meteorite swarms" and the "rings of Saturn." He obviously views these not as natural dynamic phases within our solar system, but as disrupted states of matter caused by evil spirits.

Thus we have considered Morris's research into disturbances in stars and planets. Apparently he has identified basic precepts of "creation science." Will America allow such teachings in the public schools?

Many in our society are fearful of earth invasions by evil spirits. Most fundamentalist preachers and writers foster belief in, and inspire fear of, spirits. Kelly Segraves is no exception. In his writing he is not so much concerned with physical violence among the stars and planets as he is with alleged messages from Satan delivered to earth by evil spirits. He posed a question and stated his beliefs: "What are these beings from outer space? I truly believe that we are being visited by beings that are extraterrestrial. . . . I believe that these visitors are fallen angels who have come to prepare the kingdom of antichrist on the planet."[29]

Unfortunately, we cannot expect a shortage of dedicated laborers in creationism's Surmise Quarry. Nor will creationists who build their inverted pyramid lack for the most bizarre and incredible stones cut from realms of mystery and make-believe.

Notes

1. M. Gardner, *Fads and Fallacies in the Name of Science* (New York: Dover Publications, 1957); S. Freske, "Evidence Supporting the Great Age of

the Universe," *Creation/Evolution* 2:34-39; R. Price, "The Return of the Navel, the 'Omphalos' Argument in Contemporary Creationism," *Creation/Evoltuion* 2:25-33; R. J. Schadewald, "Modern Creationism and the Ghost of Gosse," *Reports* 11(1):20-21.

2. P. Gosse, *Omphalos: An Attempt to Untie the Geological Knot* (London: John van Voorst, 1857).

3. J. C. Whitcomb, *The Early Earth* (Grand Rapids, Mich.: Baker Book House, 1986), p. 46.

4. P. A. Bartz, "Questions & Answers," *Bible Science Newsletter* 28, no. 8 (1988):5, 16.

5. J. C. Whitcomb, *The World That Perished* (Grand Rapids, Mich.: Baker Book House, 1973); J. C. Whitcomb and H. M. Morris, *The Genesis Flood* (Phillipsburg, N.J.: Presbyterian and Reformed Publishing Co., 1961).

6. R. Young, *Analytical Concordance to the Bible* (Grand Rapids, Mich.: Wm. B. Eerdmans Publishing Co., 1952), p. 211.

7. Whitcomb and Morris, *The Genesis Flood*, p. 3.

8. Young, *Analytical Concordance to the Bible*, p. 698.

9. D. N. Schramm, "The Age of the Elements," *Scientific American* 230 (1974):69-77; S. G. Brush, "Finding the Age of the Earth: By Physics or by Faith?" *Journal of Geological Education* 30 (1982):34-58; G. B. Dalrymple, "How Old is the Earth? A Reply to 'Scientific' Creationism," in *Evolutionists Confront Creationists*, eds. F. Awbrey and W. M. Thwaites (Proceedings of the 63rd Meeting, San Francisco, Pacific Div.), *AAAS* 1, part 3 (1984):66-131.

10. G. B. Dalrymple, *The Age of the Earth* (Stanford, Calif.: Stanford University Press, 1991).

11. S. West, "Notes from Earth." *Science News* 117, no. 11 (1980):174.

12. Whitcomb and Morris, *The Genesis Flood*, p. 299.

13. Ibid., p. 256.

14. H. M. Morris, *The Beginning of the World* (Denver, Colorado: Accent Books, 1977), p. 78.

15. Whitcomb and Morris, *The Genesis Flood*, p. 121.

16. W. Lammerts, quoted in ibid., p. 170.

17. Ibid., p. 326.

18. J. C. Dillow, *The Waters Above: Earth's Pre-Flood Vapor Canopy* (Chicago: Moody Press, 1982), p. 137.

19. Whatever creationists may assume about survival of plants under conditions of catastrophic flooding, I suggest that important insights could be gained by proper testing. I propose the following experimental design: (1) select plant species (land or aquatic, salt-tolerant or not). (2) Completely submerge plants in ocean saltwater at varied concentrations. (3) Impose water pressures to the equivalent of a 7,000-foot depth. (4) Conduct experiments in complete darkness, as would occur at great water depth. (5) Periodically check cellular materials for physical disruption and osmotic pressure effects, and analyze meristematic tissues for viability. (6) Record and publish results of the survival times and death of plant species.

20. P. Osmon, "Morris on the Earth's Central Cavity," *Creation/Evolution Newsletter* 6, no. 6 (1986):15.

21. Q. Williams, R. Jeanloz, J. Bass, B. Svedsen, and T. J. Ahrens, "The Melting Curve of Iron to 250 Gigapascals: A Constraint on the Temperature at Earth's Center," *Science* 236 (1987):181–82.

22. H. M. Morris, *The Remarkable Birth of Planet Earth* (San Diego, Calif.: Creation-Life Publishers, 1978), p. 67.

23. Ibid., p. 66.

24. Ibid., pp. 66–67.

25. Ibid., p. 66.

26. Ibid., pp. 66–67.

27. Ibid., p. 67.

28. Ibid., pp. 66–67.

29. K. L. Segraves, *Sons of God Return* (New York: Pyramid Books, 1977), p. 87.

3

Creationist Manipulations
Atmosphere and Earth

Canopy Location: Specifically Where?

The creationists' concept of a vapor canopy is fraught with un-certainty and contradictions. Creationists do not know how their vapor canopy was elevated above earth.[1] And, more importantly, they have no verification that the canopy ever existed.[2]

Nevertheless, Morris developed a story of primeval waters "elevated far up in the sky, above the 'firmament.' "[3] The firma-ment, he believed, corresponds to atmosphere, and the water above it existed as invisible vapor reaching far into space.

"Far into space" lacks precision, but I have tried to depict locations of the surmised canopy in figure 3. Whitcomb and Morris[4] mentioned the thermosphere, "that region above about 80 miles," as being very hot, "possibly rising to 3000° F," and that high temperature is indeed required for retention of much water vapor (fig. 3, Region A). Furthermore, they emphasized that "water vapor is substantially lighter than air."

The same authors also supposed that a tremendous quantity of vapor, at temperatures above boiling, could have been sustained "above the stratosphere, if it were somehow placed there."[5]

On the other hand, they also thought[6] it "possible that the vapor blanket could have been in the upper troposphere, *below* the stratosphere" (fig. 3, Region B). "As a matter of fact," they

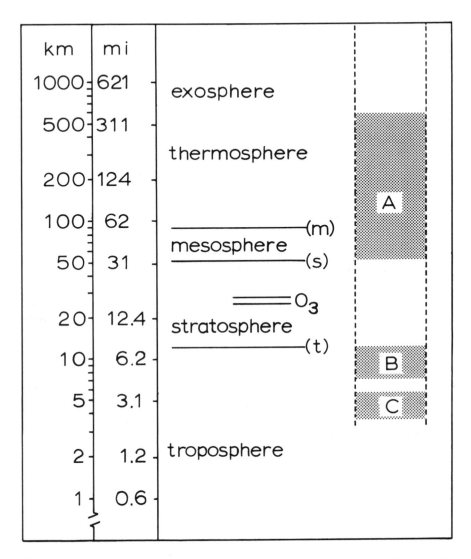

Fig. 3. Location of creationism's water vapor canopy above the primitive earth. Dashed lines indicate creationist uncertainty about vapor depth. Canopy locations (hatched areas): A. "the region above the stratosphere." B. "upper troposphere." C. "considerably below 20,000 feet." Abbreviations (pause is a boundary): t = tropopause; s = stratopause; m = mesopause. Speculations about canopy location rest on the assumption of the canopy's existence, for which creationists have no verification.

surmised,[7] "it would seem that the vapor blanket could possibly be substantially lower than 20,000 feet without being precipitated" (fig. 3, Region C).

Obviously, creationists allow themselves great latitude in defining their canopy's location. They would probably settle for any location if it were sufficiently plausible. Apparently, notable (and ostensibly scientific) authorities have not yet decided upon the most appropriate location.

Canopy Integrity: "Until Something Happened"

Creationists have no scientific evidence of their canopy's existence, how it was elevated, the quantity of water involved, or where it was located. Nor can they state what caused the alleged canopy to precipitate and fall to earth.

They maintain, without supporting evidence, that the troposphere lacked nucleating particles, due to suppressed atmospheric turbulence under a uniformly warm global climate. Whitcomb and Morris concluded, "Thus such a vapor canopy could be maintained indefinitely, until something happened to mix it with the cold gases of the stratosphere and to supply meteoric or other particles for nucleation."[8]

Such a chain of speculation is remarkable for its assumptions and its contradictions of earlier statements. Critics would like answers to important questions. Where were "cold gases of the stratosphere" located relative to the surmised water-vapor layer? How could any gas in creationism's layered atmosphere be cold? What force mixed canopy vapor with other gases?

Whitcomb and Morris proposed a canopy location "substantially lower than 20,000 feet" and a mixing of its vapor with cold gases in the stratosphere.[9] This contradicts their lighter-than-air model in which water vapor supposedly was supported above heavier gases.

The same authors suggested another location: "invisible water vapor in the region above the stratosphere."[10] If this were true, the supporting layer of gases could not remain cold. Consider the following scenario based on creationist assumptions:

1. Clouds never formed in the preflood vapor canopy.

2. Incoming Infrared (IR) energy was absorbed by the canopy.

3. All wavelengths of light penetrated the canopy.

4. Most of the light energy under the canopy was absorbed and reradiated as heat (IR) wavelengths.

5. The IR energy could not have escaped through the massive vapor barrier and returned to space.

How could cold gases exist under the creationists' vapor canopy if thermal energy had been trapped continuously under that canopy for thousands of years? Equally perplexing is how teeming life could exist under such conditions.

To summarize, for those who can believe a vapor canopy existed, it obviously existed somewhere, "if it were somehow placed there."[11] But no scientific evidence supports conjectures of how or where. And if winds could be suppressed and dust injections eliminated, the "canopy could be maintained indefinitely, until something happened."[12] But nobody knows what.

Whitcomb and Morris ended the discussion of their antediluvian vapor blanket with a bundle of guesses:

When finally that "something" happened, whatever it was— possibly the passage of earth through a meteorite swarm or the sudden extrusion of large amounts of volcanic dust into the air— the vapor blanket was condensed and precipitated.[13]

Topography: A Slight Redistribution

Only recently have scientists obtained accurate estimates of earth's water volume and mountain elevations. With such information we are now able to check biblical statements against data from earth science. We thus know that if all landforms were leveled and all ocean water were spread over a smooth globe, the water depth would be about 8,400 feet (2.6 km). If raised above sea level, the same depth of water would still be far below the peaks of earth's highest ranges.

Facing a challenge to develop a scientific story, creationists have had to deal with very high mountains and limited water. However, they remain supremely confident and answer serious problems with easy manipulations. For example, to critics who objected that not enough water existed to flood earth's highest elevations, Morris replied, "But there is plenty of water in the present oceans if the topography were slightly redistributed."[14]

Creationists present a remarkable solution to their floodwater shortage. Several things had to be done to reduce the amount required, and the first seems very easy—simply lower the mountains.

I digress here to wonder at the credulity of some who read and believe creationist fabrications. The storyteller's fantasy, conveyed in literature, recreates itself in the reader's mind and there brings down the mighty Everest and other high mountains of the world. Moreover, all ancient ranges, millions of years old, are reduced in age to less than 10,000 years.

But the mountains were not destroyed, only reduced to manageable size, according to the needs of the storyteller and the amount of water available to cover them. Whitcomb declared, "Enormously high, snow-capped mountain peaks could not have existed before the Flood."[15] And he continued with speculations about "low-lying mountains which were probably less than six or seven thousand feet high. . . ."[16]

Lowering the mountains, however, could solve only part of the floodwater shortage. Other manipulations were needed. The total "solution," as the story is finally developed, involved not only lowering mountains but uplifting ocean basins and the simultaneous collapse of the vapor canopy, which allegedly became a large part of the present ocean.

Fantastic Teeter-Totter

Creationists manipulate continental masses as well as the sea floors. Their story can be visualized as a teeter-totter or seesaw. The ocean floor is one end of a seesaw, holding ocean basins full of water. On the other end are continents with conveniently lowered mountains. Tipping the seesaw and spilling the ocean over the continents is obviously no problem for those who need only to record the "facts" on paper.

Whitcomb[17] interpreted the breakup of the "fountains of the deep" and their later closing (Genesis 7:11, 8:2). The breakup, he says, involved the "uplift of ocean floors" and the biblical reference to stopping the fountains "must refer to the reversal of this action, whereby new and much deeper ocean basins were formed to serve as vast reservoirs for the two oceans which were separated. . . . before the Flood" (Genesis 1:7).

Several questions remain unanswered. How could the ocean floors rise and remain intact—though "broken up"? And how did the ocean floor keep the oceans elevated for months before collapsing to form deeper basins?

Furthermore, what force broke up the ocean floors and raised the fractured pieces to serve as a global ocean-bearing platform? And how was that force maintained? Creationists easily tell of marvelous happenings but do not explain them. Apparently, the authors expect readers only to believe, not understand, the physical mechanisms involved.

Creationists Whitcomb and Morris proposed that several Bible verses give fundamental information about earth's geological history.[18] In the King James Authorized Version of 1611 (AV) for Psalm 104:6–9, the subject of water was discussed throughout the passage. Verse 8 had two comments about waters: they "ascend" mountains and they descend through valleys. Later versions, however, rejected those descriptions. The 1901 American Standard Version (ASV) for verse 8 provided parenthetical text: "(The mountains rose, the valleys sank down)." In the AV version the text was in a footnote. The 1952 Revised Standard Version repeated without parentheses the ASV translation that mountains rose and valleys sank. An interesting reversal in translation appeared in the New English Bible in 1970. As in the AV, water was the subject (Psalms 10:8), ". . . flowing over the hills, pouring down into the valleys. . . ."

Whitcomb discussed mountain building and asserted that the Psalmist was actually recording geological history.[19] He believed that the Psalmist's words, "the mountains rose," speak definitely about the formation of our present mountain ranges. He also stated, without regard for scientific evidence, that "Psalm 104:8 is actually saying that God supernaturally pushed up great mountain ranges in the continental areas to balance the new depths in the ocean basins."

Recognizing earth as a globe with a limited inventory of water, the creationists have readily understood the manipulations necessary to develop their flood story. First they lowered the mountains. Then they released canopy and cavern waters and, simultaneously, raised ocean floors to lift floodwater above their lowered mountains. Creationists were then faced with new challenges. They must drain floodwater from the continents. Thus, they deepened ocean basins and simultaneously elevated continents and raised mountains.

Although earth scientists have little time to argue interpretations of biblical poetry, questions about such writing deserve answers. Could "valleys" be interpreted as sea floors or ocean basins? Whitcomb and Morris obviously embraced such an interpretation in developing their story.

Answers that scientists seek are not found in interpretations of sacred writings. They ask for physical evidence and descriptions of mechanisms that could cause basins to collapse and high mountains to rise in a single year. Creationists Whitcomb and Morris[20] and Whitcomb[21] seemed satisfied in asserting that earth scientists have miserably failed to determine the driving force and mechanism responsible for mountain building. Why is it that creationists who call themselves scientists make no effort to determine and explain scientifically how mountains rose and sea floors collapsed in the latter half of their flood year?

Creationism's primary "evidence" for worldwide mountain building rests on three words, "The mountains rose," which they proclaim to be words from God. Are scientists justified in rejecting such paltry evidence as nonsense? Regardless of harsh judgement from creationists, the scientific community should—and certainly will—remain skeptical, as long as creationists embrace biblical poetry and song as their primary source for geological information about ocean basins and mountain building.

Notes

1. J. C. Whitcomb and H. M. Morris, *The Genesis Flood* (Phillipsburg, N.J.: Presbyterian and Reformed Publishing Co., 1961), p. 229.

2. Ibid., p. 241.

3. H. M. Morris, *The Beginning of the World* (Denver, Co.: Accent Books, 1977), p. 25.

4. Whitcomb and Morris, *The Genesis Flood*, pp. 240–41.

5. Ibid., p. 256.

6. Ibid., p. 257.

7. Ibid.

8. Ibid.

9. Ibid.

10. Ibid., p. 256.

11. Ibid.

12. Ibid., p. 257.

13. Ibid., p. 258.

14. Morris, *The Beginning of the World*, p. 106.

15. J. C. Whitcomb, *The World That Perished* (Grand Rapids, Mich.: Baker Book House, 1973), p. 40.

16. Ibid.

17. Ibid., p. 35.

18. Whitcomb and Morris, *The Genesis Flood*, pp. 77, 122, 267.

19. Whitcomb, *The World That Perished*, p. 39.

20. Whitcomb and Morris, *The Genesis Flood*, pp. 140–42.

21. Whitcomb, *The World That Perished*, p. 40.

4

Creationist Flood Geology

Creationism's Volcanic and Tectonic Activity

Young-earth creationists use a biblical framework for historical geology. One selected verse (Genesis 1:7) refers to elevated water in space. Another (Genesis 7:11) refers to mysterious "fountains." The combined effects of those reserves, creationists assert, are responsible for virtually all geology, especially flood geology. To understand what religious fundamentalists have taught (or have been taught) for the past thirty years, we must understand the definition of volcanic and tectonic geology. In geological science, tectonic refers to building and deformation of planetary crusts and to the forces involved; earth tectonics refers to earth's structural makeup, particularly faulting, folding, and other features of crust. The creationist version of tectonics is a distortion of legitimate earth science.

Whitcomb and Morris published a string of bold assumptions that are the base for creationist teachings on flood geology. Their geology is divided into two periods.[1]

Period 1. A quiet preflood era. The alleged preflood interval between the biblical creation and global flood was a quiescent period meteorologically, with no rainfall before the flood. Air turbulence was "largely absent" and "large movements of air masses were prevented." Further, a "probably small" cooling of air with height caused condensed vapor to fall "as a light mist soon after its evaporation" from an "intricate network of 'seas' " and languid rivers. In addition to rejecting a dynamic meteorology, creationists disavow intense volcanic and tectonic activity. The absence of crustal change in the imagined preflood era contrasts sharply with events of creationism's flood year.

Period 2. A breakup of the "fountains of the deep." The "deep" apparently has been a very mysterious term for creationists and has had several meanings. It allegedly referred to ocean water and water elevated into a vapor canopy, or trapped underground water, or "magma and water or steam." In Whitcomb's view, the breakup of the "fountains of the deep" refers to the tectonic "uplift of ocean floors."[2] Water "imprisoned" during the preflood time, "perhaps steadily building up temperatures and pressures," finally escaped. The explosive release of trapped liquids plus collapse of the vapor canopy allegedly caused "further fractures" and "earth movements" and virtually all geological and tectonic effects. Obviously, ICR creationists want to keep all massive movements of water and land masses within their flood year. But they also want to allow some "residual catastrophism" as decaying flood effects, such as volcanic activity, mountain building, "and possibly continental drift. . . ."[3]

Creationism's Pressure-Cooker Geology

The creationists' preflood world can be divided into two compartments: outer and inner reservoirs of energy. The outer compartment included earth's surface and atmosphere, with a very thick water-vapor cover that absorbed solar heat and transmitted light energy to the lower atmosphere and earth. That compartment was not coupled to earth's internal energy.

The inner compartment was a sealed pressure cooker with its own internal energy (heat of compressed matter and radionuclide decay). That compartment was "postulated" as "vast subterranean heated and pressurized reservoirs, perhaps in the primeval crust or perhaps in the earth's mantle itself. . . ."[4] Earlier, Whitcomb and Morris described the explosion of their pressure cooker: trapped water, "perhaps steadily building up temperatures and pressures until . . . the crust gave way" led to universal fracturing and breakup of "all the fountains of the great deep. . . ."[5]

For creationists, this was a major event. As catastrophists,[6] they often like to emphasize that all their "pressurized reservoirs" were involved. Their conjectures required a sealed and pressurized vessel that could store up energy and then burst open to do virtually all of earth's geological work in one year.

The pressure cooker disintegrated, and creationists tell us much about the resulting pieces—dust, magma, crustal slabs, and so on. In the alleged flood year the pieces fell back and were arranged in new structures, such as global sediments, high mountain ranges, deepened ocean basins, and volcanic islands. Geologists, of course, reject such recent, rapid, and explosive scenarios.

Creationists counter with the remarkable quibble that their opponents' uniformitarian model has always lacked an energy source for the tremendous geological events on earth.

Scientists must claim, of course, that sufficient energy has always been available to account for earth's geological activity. Moreover, much of that energy, gauged by earth's changing surface features, was released slowly—certainly not in a universal explosion with a total reconstruction in one year.

Scientists cannot accept creationism's sealed pressure cooker as analogous to the structure and dynamic behavior of earth's crust and internal matter. In the first place, the crust has not been a tightly sealed cover. Through millions of years, the crust has vented gases and magma from mid-ocean ridges, from volcanoes of numerous island arcs, and from hotspots on the ocean floor. The world's ocean floors, volcanic mountains, and seamounts are tectonic monuments that remind us of both ancient and present openings to earth's interior.

One of the most complete and readily available depictions of earth's tectonic activity during the past 20 million years is the detailed map published in *National Geographic* magazine.[7]

Deep and Shallow Sediments on Ocean Floors

Creationist views differ drastically from modern scientific views on sediment distribution over ocean floors. I propose three scenarios and related sediment patterns. Two of them I ascribe to creationists, and the third to geologists.

1. Global erosion that caused worldwide transport of sediment took place about 4,350 years ago in a universal flood, and no

ocean plate has moved since that time. Because massive quantities of materials were deposited over stationary ocean floors, sediments should extend over the floors in layers of relatively uniform depth.

2. Global erosion and transport of materials occurred as stated above, but an ocean plate, under a heavy load of flood sediment, moved away from a mid-ocean spreading ridge at the rate of 2.5 centimeters (1 inch) per year for 4,350 years. Because the plate moved after amounts of flood-deposited sediments were laid down, the deposits would be shallow near the spreading center and deepest away from the center—beyond 100 meters (4,350 years × 2.5 cm = 108 meters or 354 feet).[8]

3. No global flood occurred, but natural deposits accumulated on an ocean plate that moved from a spreading ridge 2.5 centimeters per year for one million years. After a million years, a substantial sediment thickness on plates moving 2.5 centimeters per year could be expected at 25 km (15 miles) from the spreading center.

Which of the three statements will the reader choose? Creationists who reject ocean-floor spreading might reasonably accept 1 as true. Others might prefer 2. Most scientists, I think, would accept 3 as one that can be confirmed by direct measurement.

Any statement selected from the above choices is meaningless if it does not correspond to a real situation. Current ocean-floor research brings the above scenarios into better focus.

Scientists and drill teams have faced serious problems when they have tried to study active volcanic, tectonic, and hydrothermal processes along ridge crests. Stefi Weisburd explained:

> But drilling into volcanic rocks younger than a million years old has been very difficult. Young crust isn't covered by the thick layers of sediments needed by traditional drilling techniques to stabilize the drill string before the hole is spudded [prepared for drilling]. And young volcanic rocks are highly fractured and abrasive, destroying drill bits rapidly.[9]

The undeniable fact is that young crust does not have thick layers of sediments. The difficulty of drilling into "rocks younger than a million years" raises serious concerns about creationist assumptions.

Creationists need to answer this question: Why is it that drillers in mid-ocean have moved back from spreading ridges to sediments that cover million-year-old sea floor? If the biblical flood story were true, the deep sediment should be near instead of miles away from the spreading centers. Scientific evidence exposes the flimsy base on which creationists try to establish a flood geology that occurred within the past 4,500 years.

Seamounts: Creationism's "Drowned Islands"

Creationists manipulate many things: mountains, oceans, ocean floors, and islands. To this list, we must add submarine mountains, called "seamounts" (or "guyots"). Seamounts are sometimes isolated features but often occur as a chain; for example, the Emperor Seamount formation in the northern Pacific.

Whitcomb and Morris declared that " 'seamounts' . . . are nothing more than drowned islands out in the middle of the ocean."[10] They defined seamounts as ". . . flat-topped, and therefore non-volcanic, and are now in many cases more than 1,000 fathoms [over 6,000 feet] below the surface."[11] These authors quote rather extensively from the respected marine geologist Edwin L. Hamilton.[12] But in reaching their conclusion, they chose to contradict Hamilton and publish their arbitrary notion that seamounts are "drowned islands" and are "non-volcanic."

Hamilton discussed fossils of coral and foraminifera obtained from drilling activities in the central Pacific.[13] The projects were at Eniwetok and Bikini, and on a "flat-topped seamount . . . (Sylvania Guyot)." He determined the age of the various fossils and sediments to be in the Oligocene and Eocene epochs. Hamilton then summarized: "These datings, plus the geophysical work at Jaluit and in the Marshall Islands, gives definite proof that the coral atolls of this area are calcareous caps on great volcanic mountains—mountains which were formerly at, or near, the sea surface when the coral began to grow."[14] The "great volcanic mountains" refute the creationist notion of "flat-topped . . . non-volcanic" formations.

Hamilton indicated that the volcanic mountains were flat-topped because of the widespread presence of "pyroclastic material (ash, lapilli, and so on) thrown out by some of the Pacific islands."[15] *Pyroclastics* are materials ejected from volcanoes. *Lapilli* are materials from 2 to 6 mm diameter. The specific reason Hamilton gave for the flat tops was that such materials erode quickly to form a submerged bank where coral can grow. In his judgment,

> The coral atoll problem . . . has been solved. These giant coral structures [in the Central Pacific] are caps on the tops of submarine volcanoes . . . and many of them were apparently planed flat at the sea surface, where coral began to grow. Later, the coral-capped seamount subsided slowly beneath the surface, and the coral kept pace with the sinking to form the atolls we see today.[16]

Why would Whitcomb and Morris take a position exactly opposite to that of Hamilton, the widely recognized expert in seafloor geology? Did they willfully sidestep geologic reality to assert an utterly false conclusion, one that nevertheless would ostensibly support the story they were developing? Probably only the authors can disclose their motivations. Quite likely, the scientific evidence reviewed by Hamilton conflicted with notions of the flood story creationists were constructing. Thus, creationists must reject millions of years of seafloor subsidence to defend Bible chronology.

Hamilton opened the door to the subsidence question by mentioning "a small minority" who believed that ocean water increase, not sinking seafloor, could account for the present underwater location of seamounts.[17] Hamilton further speculated that if water volume had increased, then "something on the order of a 30 percent increase in the volume of the oceans must have occurred during the last 100 million years."[18] That statement seems to be exactly the one Whitcomb and Morris were looking for—except that they must reduce the "100 million years" to one flood year, occurring a few thousand years ago.

Thus, unwittingly, Hamilton suggested a formula that filled creationist needs. They responded:

> And if the second alternative is chosen, that of a relatively sudden [interpret: sudden means biblical 40-day] increase of 30 per cent in the volume of the ocean, the compelling question of the source of this water must be faced, and this few geologists can bring themselves to do![19]

The last sentence of the quote deserves rebuttal. Obviously Whitcomb and Morris consider themselves honest and objective. But what do they imply about most geologists? Do professional geologists accept, without verifiable data, a sudden 30 percent increase in ocean volume? Further, many others in the scientific community cannot bring themselves to forfeit rationality in believing absurd details in the biblical flood story. Chapter 12 explains how Whitcomb and Morris altered a basic part of the biblical flood story—a part which they, apparently, could not themselves believe.

Whitcomb and Morris concluded that the problem about the source of the water that covered their seamount islands "becomes simple" if people believe that "waters above the firmament" precipitated in the biblical flood.[20] They stated the issue correctly. For them, water that covered seamounts was not a problem resolved by scientific research, but a matter settled by religious faith and belief.

Seamounts continue to rise from deep ocean floors. In October 1987, oceanographer Harmon Craig observed and took samples of glassy basaltic materials erupting from MacDonald seamount in the South Pacific near French Polynesia.[21] This volcano is at the end of a 2,000-km (1200 mile) chain of islands and seamounts. Scientists concluded that the chain was built from molten rock one link at a time over millions of years as the Pacific plate moved over a "hotspot" open to the earth's interior. The growing seamount was 5,000 meters above the ocean floor and only 40 meters below the ocean surface. Monastersky stated that this chance observation should "help scientists understand the evolution of seamounts, which often become islands." The Hawaiian Islands are an excellent example of such evolution.

Seamounts are being built up in ways other than by hot basaltic lava.[22] The Ocean Drilling Program took seamount samples composed of serpentinite, a nonvolcanic rock, which is a mixture of water and minerals from the earth's mantle, from near the Mariana Trench, south of Japan. The operation yielded soft serpentinite "mud" from the interior and also near the top of Mariana seamount, more than a mile above the subduction zone, where the Pacific plate moves under the Philippine plate.

The structure of many atolls and the composition of volcanic and nonvolcanic seamounts show that seamounts have built up over very long periods. The direct observation of magma releases demonstrated that seamounts are still being lifted above ocean floors. We conclude that such structures were not created as islands at an alleged creation a mere 10,000 years ago. Neither did islands become seamounts in a conjectured flood year 4,300 years ago.

The Biblical Flood: Universal Threat to Life

Flood geologists repeatedly emphasize the role of flood water in the worldwide distribution of sediments. They contend that ocean water plus water from their vapor canopy and the ruptured earth, taken together, would have equaled the quantity in the present oceans—about 316 million cubic miles or 1.32 billion cubic kilometers.

What is the estimated mass and volume of sediment now existing on earth? Creationists assume that essentially all ocean and dry-land sediments are those distributed by the Genesis flood water.

"Most recent estimates give values for the total mass of sediments of about 2.5×10^{24} g, equivalent to about 10^9 km^3 of sediment or an average thickness over both continents and oceans of about 2 km."[23]

If the nearly equal volumes of water and sediment were mixed, they would form a slurry in which water was only slightly in excess of sediment: 1.32 billion cubic kilometers of water plus 1 billion cubic kilometers of sediment.

Compared to the 2-km shell (more than 6,000 feet) of sediment now covering the globe, the sediment accumulation that creationists would allow in their entire preflood era was almost nil. They assumed a quiet, placid world: gentle springs, tranquil rivers, no winds, rainstorms, volcanic emissions, and so forth, and, therefore, no erosion or other processes that might produce sediments.

Whitcomb and Morris speculated extensively about processes and effects of the alleged Genesis flood:

> *Volcanic and Seismic Upheavals* . . . Great volcanic explosions . . . great quantities of liquids, perhaps liquid rocks or magmas . . . water (probably steam) . . . under great pressure

... burst forth ... probably both on the lands and under the seas. ... [T]here must have been great earthquakes and tsunamis ... generated throughout the world.

Unprecedented Sedimentary Action ... tremendous quantities of earth and rock ... excavated. ... Many factors ... driving rains ... raging streams ... earthquakes and volcanic eruptions ... powerful tidal waves, then later the rising of the lands and sinking of the basins, and perhaps many other factors which we cannot now even guess.[24]

After all, guessing takes time. Later, we read that, "Great tidal waves undoubtedly were generated in prodigious numbers, as the imprisoned waters progressively escaped through crustal fractures all around the earth. ..."[25] And, "Both the rains and upheavals apparently continued for at least 150 days."[26]

Whitcomb and Morris have tried to orchestrate the ancient burial and fossilization of marine and other animals. And, of course, they related all this to the biblical flood.

They began at the ocean floor:

The creatures of the deep sea bottoms would universally be overwhelmed by the toxicity and violence of the volcanic emanations and the bottom currents generated thereby and would in general be mixed with the inorganic materials. ...[27]

This statement seems to indicate the total extinction of marine organisms.

In similar fashion, the fish and other organisms living nearer the surface would subsequently be entrapped by either materials washing down from the land surface or the shallow coastal sea bottoms or by materials upwelling from the depths.[28]

Again the result was entrapment and extinction.

Land animals and man were assumed to be more mobile than others, temporarily escaping to higher ground before dying in the overflowing water.[29] The alleged kinds of animals on Noah's ark were to replenish land animals on earth. No marine or freshwater forms were specified.

Thus, creationists believe they have constructed an impregnable framework for one segment of world history. They also believe that they have confounded the uniformitarians by describing how simple and complex forms of life were overwhelmed in deep sediments to become fossils—not over immense geological ages but in the first few days or perhaps hours of a universal flood.

Creationists have loosed such a deluge of words describing the flood that one might wonder how any organisms survived. Consider the ramifications of their fictional story.

The watery environment was essentially a 1:1 mix of saltwater and sediment. This dangerous and universal slurry was washed back and forth across the globe by "Powerful currents, of all directions and magnitudes and periods. . . ."[30] The slurry overflowed the highest mountains and was continuously stirred laterally by currents and vertically by volcanic explosions "for at least 150 days." The "toxicity and violence of volcanic emanations" could have destroyed life by the impact of repeated explosions, and the toxic volcanic gases mixed with the slurry of mud would have blocked respiratory processes in all water-dwelling organisms.

Somehow, creationists assumed that bottom-dwelling organisms and the fish and other forms that inhabited upper levels of marine and fresh water survived months of deadly upheaval.

In their eagerness to validate the biblical narrative and to confound their opponents, creationists have proved too much. Extinction is forever. They should cease writing about annihilations and explain how present forms originated.

Will Modern Creationists Embrace Plate Tectonics?

Laurence Gould, in a foreword to the book *Wandering Lands and Animals,* stated that plate tectonics theory

> has created the greatest revolution in man's thinking about his earth since the Copernican Revolution. . . . This theory has simplified and unified all the main fields in the earth sciences such as paleontology, paleobotany, sedimentary geology, ore deposits, seismology and volcanology.[31]

Religious conservatives and others who perhaps may understand certain ramifications of plate tectonic theory will, nevertheless, ignore that unifying concept. And in its place they will offer pretensions of religious myth.

Thirteen years after *The Genesis Flood* was published, the Institute for Creation Research (ICR) produced its most "authoritative" statements of policy and beliefs in the book *Scientific Creationism*.[32] It stated:

> The creation model makes no specific predictions regarding continental drift, so is not affected either way. However, one of the main difficulties with the concept as developed in a uniformitarian context has always been the absence of a source for the tremendous energy required to drive continents apart. The cataclysmic model, with its tremendous store of subterranean energy suddenly released at the time of the Flood, alone seems capable of accounting for the energy.[33]

Seemingly, Henry Morris and the ICR staff thought it advantageous to sit on the fence and hope that their creation model would not be "affected either way" by their ostensible neutrality. They did not blast continental drift as totally impossible—only that the uniformist model had always lacked the energy to move continents. Obviously, these creationists could not publicly recognize natural and relatively slow-working geological processes. Such an act would reveal them as uniformists and evolutionists and would compromise their recent, explosive, pressure-cooker geology.

How long will creationists continue to embrace the fiction spelled out by Whitcomb and Morris in 1961? For example, on the subject of tectonics and mountain building in geologically recent times (Pleistocene or early Pliocene) these authors mentioned "folds, faults, rifts, thrusts" as being active.[34] They then came to this incredible conclusion: "*But they are not active now*, at least not measurably so!" (Whitcomb and Morris's italics.)

Scientific studies are currently measuring the activity and movements of faults, thrusts, and other related phenomena. Consider the following examples.

The San Andreas is a well-known example of an active crustal fault. Displacements in the 1906 earthquake measured up to 6 meters (20 feet) along a section of the fault on the Point Reyes

Peninsula. Measurable plate movements along active fault lines occurred in the 1989 Loma Prieta earthquake.[35] Rates of slippage along many other faults have been recorded over the past decade in the scientific literature.

Rates of seafloor creation can be measured. Cooling magma of new seafloor acquires a magnetic polarity that aligns with earth's polarity. A magnetic stripe remains in the floor and generally parallels the spreading ridge. The magnetic signature offers investigators a gauge for the rate of seafloor creation and a reliable map of seafloor movement. Smith provided a diagram of a standard polarity sequence for the South Atlantic Ocean basin for the past 65 million years.[36]

Rifts in the ocean's crust are still occurring. Ancient Pangaea separated into the continents of Africa, South America, India, Australia, and Antarctica. These lands then drifted to their present locations and are now separated thousands of miles by oceans created by ocean floor spreading.[37]

The Indian plate slid into the Eurasian plate millions of years ago. Its continuous force accounts for the greatest mountain building on earth, as faults and massive folding push up high plateaus and the Himalayan range.[38] Tibet and India converge at the rate of about 6 centimeters per year. The piling up and thickening of underlying crust, along with fault activity and folding, continuously shorten the distance between the plates. Mount St. Elias, an 18,000-foot (5,486 meter) peak in Alaska, is being pushed up 5 centimeters (2 inches) per year by thrusts from the Pacific Plate.[39] Garrett depicted approximately 46,000 miles of mid-ocean ridges and gave rates (in millimeters) at which seafloor spreads from ridges.[40] This map identified more than twenty hotspots and gave the rate and direction of ocean plates passing over them.

Continents are still rifting.[41] In East Africa, stretched crust is dropping along fault lines and creating rift valleys. The valleys extend from the Red Sea more than 620 miles (1,000 km) through Ethiopia and Kenya and into Tanzania. Another rift system stretches more than 930 miles (1,500 km) from southern Sudan around Lake Victoria and south into Malawi. Rifting in the Gulf of Aden has already created oceanic crust, and an oceanic rift has developed and is widening in the lower Red Sea. Relatively new oceanic crust, with embedded magnetic stripes, is also forming in several linear sections of the central Red Sea.[42]

Plates that have transported continents are still moving. For example, on the IndoAustralian Plate, seafloor is spreading from the Mid-Ocean Ridge and carrying its recorded lines of polarity that provide a measure of time and the distance over which the Australian and Antarctic continents have separated. Another indicator of time and distance of plate movement is a biological one—the Great Barrier Reef. The reef stretches 1,240 miles (2,000 km) along Australia's northeastern coast and is now between 9°S and 24°S latitude. Seismic and drilling tests have revealed much thicker coral growth in the northern than in the central and southern sections of the barrier. Research indicates that plate migration carried successive regions of the Great Barrier Reef out of the temperate and into the tropical zone. The northern coral reefs made that transition between 16 and 25 million years ago. Transition to the tropical environment by the central reefs occurred between 10 and 15 million years ago. The southern reefs have been similarly exposed for only the past few million years. The entire reef now develops in tropical water where surface temperatures (annual mean) are generally above 20°C.[43]

Quite clearly earth plates move. The ICR's notion of a no-energy-source for continental drift is a perfect quibble. Apparently, creationists can do nothing more than indulge in temporizing. They gain time while they sit on the fence and put off decisions. Theatrics on the fence, however, may become increasingly difficult, particularly if sitters are aware of barbs. Creationists, of course, are dreadfully aware of such barbed concepts as the origin of new forms by natural processes. Seafloor spreading slowly creates new floor, and new ocean. An ocean can also be eliminated when a continental plate pushes into another, for instance, when India approached and contacted mainland Asia. Creationists apparently could tolerate such facts of ancient and modern history if they could cram all (or most) of earth's geology into one biblical year.

The ICR has an imaginative plan for compressing geologic time into flood sequences.[44] The goal for their model "would be to organize the geologic strata of the earth into a standard geologic column" that would conform to stages of their imagined flood. They suggest an equation that converts scientific geologic ages into chronologic units corresponding to their flood stages. For example, the Archeozoic would correspond to "Origin of crust dating from the Creation Period. . . ." Mesozoic would refer to intermediate flood phases, and Pleistocene would denote a "Post-Flood" time involving much less tectonic activity. Contemplating the large effort needed "to work out the details" of their proposed revision, the ICR registered a call for teachers to encourage gifted students to seek careers with such a goal in mind.[45] Library research and some narrowly prescribed field work would likely be entailed in such careers. Resolving chronological difficulties that exist between the Bible and modern science would remain as the first priority.

Young-earth creationists are not ready to accept plate tectonics as a unifying concept of earth sciences. At most, they might agree that continents could have collided and raised up mountains as they skittered over the globe in the flood period. They give a "perhaps" statement about such activity as their best explanation for mountain-building:

> The vast isostatic readjustments necessary after the Flood, perhaps augmented by drifting and colliding continents also triggered by the Flood, provides the best explanation of mountain-building now available.[46]

Creationists, particularly ICR personnel, do not want to face the critical issue in continental drift. The important concern is not an imagined quantity of stored energy that was suddenly released in short-time, explosive, pressure-cooker geology. The real issue is the potential for slow release of earth's internal energy, made available at its surface for longterm phases of construction and destruction of ocean plates, mountains, and other tectonic features.

Notes

1. J. C. Whitcomb and H. M. Morris, *The Genesis Flood* (Phillipsburg, N.J.: Presbyterian and Reformed Publishing Co., 1961), pp. 241–43.

2. J. C. Whitcomb, *The World That Perished* (Grand Rapids, Mich.: Baker Book House, 1973), p. 35.

3. H. M. Morris (ed.), *Scientific Creationism,* General edition (El Cajon, Calif.: Master Books, 1974), p. 128.

4. Ibid., p. 125.

5. Whitcomb and Morris, *The Genesis Flood,* p. 242.

6. Creationist catastrophists hold that forces operating in the past—but not observed today—suddenly brought about global changes in the earth's crust. Modern uniformitarians recognize that catastrophes such as earthquakes, meteoric bombardment, volcanic explosions, etc., occurred and continue to occur. These events could have been more frequent on the primitive earth, but scientists maintain that the same physical laws and processes operating in the past operate today and can account for geological change. Thus, earth-building processes have mostly been uniform throughout earth's history.

7. W. E. Garrett (ed.), "Earth's Dynamic Crust," *National Geographic* 168, no. 2 (August 1985).

8. Natural deposition rates over the 4,350 years would account for sediments progressively deeper at sites near continents, the source of eroded matter. Creationists, however, believe that large sedimentary depostis in the oceans—often far from continents and thousands of feet deep—are almost exclusively the work of a universal flood.

9. S. Weisburd, "Leg 106 Treats: Hot Vents, Sea Creatures, Engineering Feats," *Science News* 129, no. 4 (1986):54.

10. Whitcomb and Morris, *The Genesis Flood*, p. 124.

11. Ibid., p. 125.

12. E. L. Hamilton, "The Last Geographic Frontier: The Sea Floor," *Scientific Monthly* 85 (1957):294–314.

13. Ibid., p. 305.

14. Ibid.

15. Ibid.

16. Ibid.

17. Ibid.

18. Ibid.

19. Whitcomb and Morris, *The Genesis Flood*, p. 326.

20. Ibid.

21. R. Monastersky, "Seamount Serendipity in the South Pacific," *Science News* 132 (1987):262.

22. R. Monastersky, "Where Earth's Insides Ooze Out," *Science News* 136 (1989):15.

23. H. Blatt, G. Middleton, and R. Murray, *Origin of Sedimentary Rocks*, 2d ed. (Englewood Cliffs, N.J.: Prentice-Hall, 1980), p. 34.

24. Whitcomb and Morris, *The Genesis Flood*, pp. 122–23.

25. Ibid., p. 261.

26. Ibid., pp. 265n, 127.

27. Ibid., p. 265.

28. Ibid.

29. Ibid., p. 266.

30. Ibid., p. 265.

31. E. H. Colbert, *Wandering Lands and Animals* (New York: E. P. Dutton, 1973), pp. xv, xvi.

32. Morris, *Scientific Creationism*, p. 128.

33. Ibid.

34. Whitcomb and Morris, *The Genesis Flood*, p. 142.

35. P. Segall and M. Lisowski, "Surface Displacements in the 1906 San Francisco and 1989 Loma Prieta Earthquakes," *Science* 250 (1990):1241–44.

36. D. G. Smith (ed.), *The Cambridge Encyclopedia of Earth Sciences* (New York: Crown Publishers/Cambridge University Press, 1981), p. 170.

37. Colbert, *Wandering Lands and Animals*, pp. xix, xxi.

38. Garrett, "Earth's Dynamic Crust." See detailed map.

39. R. Gore, "Our Restless Planet Earth," *National Geographic* 168, no. 2 (1985):158.

40. Garrett, "Earth's Dynamic Crust." See detailed map.

41. E. Bonati, "The Rifting of Continents," *Scientific American* 256, no. 3 (1987):96–103.

42. Ibid., p. 99.

43. P. J. Davies, P. A. Symonds, D. A. Feary, and C. J. Pigram, "Horizontal Plate Motion: A Key Allocyclic Factor in the Evolution of the Great Barrier Reef," *Science* 138 (1987):1697–1700.

44. Morris, *Scientific Creationism*, p. 129.

45. Ibid., p. 130.

46. Ibid., p. 126.

5

Aspects of Creationist Plant History

Spurious Age: Things Young Are Old

The creationists have a catchphrase called "appearance of age" which they apply to entities at the moment of their alleged creation, for example, a man or a tree. For them, created things only one second old would display evidence of maturity—and spurious age.

John Whitcomb declared that fruit trees did not evolve but instead, ". . . were created full-grown (without growth rings)."[1] Neither did humans evolve, he says, but "were created full-grown (without navels). . . ."

Bartz added his own ideas about appearance of age.[2] He suggested that created trees probably had ". . . rings in them just as the sea water may have had elements dissolved in it. . . ." Apparently he was not quite ready to disclose that God took up hydroponics culture and made the ocean a fertile nutrient medium for living things. However, we do learn from Bartz that God faced a critical soil problem and took up organic gardening as a necessary part of creation.

Plant Nutrition, and God the Organic Gardener

The young-earth creationists make no serious defense of the biblical statement that plants were established before sunlight (Genesis 1:11, 14). Nor are they concerned about the false notion that the first living things on earth were seed plants and fruit trees with seeds in the fruit (Genesis 1:11, 12).[3] But they do worry about the soil in which the alleged first plants grew.

The soil problem, as Bartz apparently imagined it,[4] was as follows. First, he believed the biblical record (Genesis 1:9–11) that dry land and mature plants were created on the same day. Second, he knew that organic residues accumulate in soils over many years, and he believed that organics were always necessary to generate fertile soils. But because a few hours in one creation day gave too little time, what could be done to short-cut the process? He decided that synthetic organic matter needed to be added to the soil. Imagine the scenario: God must create and apply to the ground a unique fertilizer of organics. Bartz described it thus: "The soil, in order to support life, had organics in it from things which never had lived."[5]

The Bartz story illustrates how some creationists find gaps in Bible stories and feel compelled to fill them with unsubstantiated folklore. Regardless of his speculation, soils need not contain organic matter before they can support plant life.

Organic matter in soil has two effects. It improves soil texture; for example, it makes a heavy clay soil more friable. Also, residues tend to improve water retention, such as in a porous, sandy soil, and thus prevent rapid loss of water from root zones.

The "organics" surmise of creationists is easy to refute. The following examples are instructive. On Lanzarote in the Canary Islands, grapevines grow not in organic soil but in loose granules of lava cinders.[6] For successful vineyard growth, the cinders must retain sufficient moisture from intermittent rains. Also, dwarf buckwheat plants carpet the cinder landscape of the Craters of the Moon National Monument in Idaho.[7] In the newly cultivated desert in the southwestern United States, irrigated sand that has no developed organic layer is remarkably productive—without organic supplements. Also, lichens grow worldwide not on organic soils but on bare rocks.

A number of mineral elements serve as essential nutrients when absorbed and translocated by plants. Plants assimilate *mineral elements* in processes that involve *water*, metabolic *gases*, specific energies of *light*, and ambient *thermal energy* at physiological temperature, and thus create biomass that may serve as food at other trophic (feeding) levels. All of the above factors (italicized) are necessary for most of the surface life on the planet. Some organisms, shielded from solar energy in hot springs or near deep-sea vents, use chemical energy from metal sulfides and thermal energy supplied by heated vent waters.

Numerous plants display nutritional independence of soil. They can be grown to maturity with nutrient solutions, without additions of organic agents. Such plants are called *autophytes*, which means that from simple *inorganic* substances they can synthesize their own food. The more specific term, *autotrophic* (self-nourishing), refers to green plants and certain bacteria that produce carbohydrates by capturing the sun's energy in light-sensitive pigments. Such plants do not require organic nutrients. They use carbon dioxide or carbonates as a sole source of carbon along with simple inorganic nitrogen for synthetic processes.[8]

Other plants, such as molds and mushrooms (the fungi), are defined as *heterotrophic*, meaning that they lack true chlorophyll and therefore depend on outside sources for complex organic compounds to obtain carbon and nitrogen. *Parasites* live on or in other organisms and obtain all or part of their required organic nutrients from their host. *Saprophytes* live on dead or decaying organic matter. Most plants are not parasitic or saprophytic; they are autotrophic.

Folklore abounds concerning the nutritional needs of plants and animals. Such information may contain a strong mix of religious ingredients. Kelly Segraves, president of the Creation-Science Research Center (San Diego) asserted that,

> Before the Flood there were tremendous amounts of minerals and trace elements in the soil necessary to produce the vitamins and proteins needed for humans to gain all the nutrition necessary for life. Animals shared this same food.[9]

Segraves also stated: "After the Flood these trace minerals no longer remained in the soil, but were washed out to sea."[10] That notion is manifestly false. If it were true, earth's land-dwelling organisms—including Segraves—would not exist. Moreover, he erroneously speculated, "This [trace minerals washed out to sea] may account for the fact that the largest animal living today, the whale, exists in the sea."

Segraves continued his speculations in a subject area which, through incessant propaganda, now represents a nationwide scam: vitamin and nutrition therapy. He would have us believe that

> Today man needs to take supplementary vitamins because there is not enough protein or minerals in the soil to enrich the vegetation. Before the Flood the earth had a tropical climate and mineral-enriched vegetation which were quite conducive to producing large forms of life.[11]

An excellent antidote for the unscientific organic and mineral supplement propaganda one hears today is the book *Health Quackery*.[12]

Curses on "Basic Physical Elements . . . and All Flesh"

The Bible (Genesis 1–8) describes three events that creationists consider fundamental to their science: a recent creation, a sin of disobedience (the "fall of Adam"), and a universal flood. The doctrine of the fall has special importance for creationists. That event, they believe, carried innocent humanity across a sin boundary and brought curses from God. Creationists assume that before the fall no death or any occasion for injury and suffering was possible. They believe that after the fall natural death, suffering, killing, and meat eating (subjects taken up in chapter 6) became reality.

Creationists believe that the curse also involved plants. They take literally the statement (Genesis 3:17–18) that the ground cursed by God brought forth thorns and thistles. Obviously, that is a matter of faith. Science cannot verify that the first thorny growth on earth resulted from a curse.

What information can creationists produce concerning altered plant structure that turned some plants into death traps? If death and decay did not exist before the fall, then ants did not die in pitcher plants, nor did any plant ever trap an insect and extract its body fluids. In the creationist view was the sundew a benign species on the day Adam sinned, but transformed the next day into a plant with sensitized tentacles that could close down on captive insects?

The ICR staff apparently have wished to identify exactly what was cursed, and the effects. They asserted that by cursing the ground God cursed the "basic physical elements . . . and all flesh constructed from those elements was also cursed."[13] Obviously, Henry Morris and the ICR staff cannot explain in scientific terms what they have claimed to be fact.

Unique structures that allow plants to capture insects are only part of a very complex ecology. Insectivorous plants, such as the bladderwort and pitcher plants, excrete chemicals that digest their insect prey.[14] The bladderwort, a submerged aquatic plant, has many small water-filled bladders (modified leaves). Small animals swim into those chambers via trapdoors and in time are digested. Did creationism's God create the trapdoors and perfect the chemistry of digestion? Or do creationists fix blame on cursed elements of the soil?

Processes that dissolve insect parts, then translocate the chemical constituents, followed by their absorption and, finally, their assimilation at metabolic sites, point to a very complex physiology of plant nutrition. Were an angry God and cursed elements responsible for such processes?

Creation pseudoscience presents biblical myth and allegory as literal fact. Creationists have not offered scientific data to show that radical changes in plant physiology took place after the purported fall—creationism's sin boundary. Nor have they found evidence in the fossil record that radical structural changes in plants occurred within the last 10,000 years.

Should creationist beliefs and teachings be accepted and their textual materials be certified for use in America's public schools? What might the public school curriculum be if evangelical creationists had authority to prepare Bible-based courses in plant history and science?

Would teachers of botanical science tell students that the first organisms on earth were flowering plants (Genesis 1:11) and that they were established before the origin of the sun (Genesis 1:16)? Would part of the history of plants be traced to a biblical snake fable and be presented to children as absolute fact? Would instructors in botany teach that drastic transformations of plants were caused by cursed mineral elements in soil? Would instructors in the life sciences teach that trophic (feeding) levels never existed before Adam's sin and that all marine and land animals ate vegetation, and not a single food chain or food web existed on earth? American education faces serious threats to the integrity of the system. Educators concerned with biological and other teaching programs have a crucial responsibility in establishing curricula that preserve and promote quality education in American schools.

Notes

1. J. C. Whitcomb, *The Early Earth* (Grand Rapids, Mich.: Baker Book House, 1986), p. 46.

2. P. A. Bartz, "Questions and Answers," *Bible Science Newsletter* 28, no. 8 (1988):5.

3. Biblical writers were not aware that ancient plants thrived in the Paleozoic Era millions of years before flowering plants began to flourish in the Cretaceous Period (G. G. Simpson, *Fossils and the History of Life* [New York: Scientific American Books, 1983], pp. 59, 82).

4. Bartz, "Questions and Answers," p. 5.

5. Ibid.

6. R. D. Ballard, *Exploring Our Living Planet* (Washington, D.C.: National Geographic Society, 1988), p. 184.

7. Ibid., pp. 202, 203.

8. Roots of autotrophic plants also have the ability to extract trace elements bound to organic molecules in rooting media. In demonstrations in controlled cultures, roots separated iron from highly stable organic metal chelates and left the organic agents in the rooting medium (L. O. Tiffin and J. C. Brown, "Absorption of Iron from Iron Chelate by Sunflower Roots," *Science* 130 [3370]: 274-75; L. O. Tiffin, "Iron Translocation I. Plant Culture, Exudate Sampling, Iron Citrate Analysis and II. Citrate/Iron Ratios in Plant Stem Exudates," in *Papers in Plant Physiology,* ed. W. S. Hillman [New York: Holt, Rinehart, and Winston, 1970], pp. 49-57; and L. O. Tiffin, "Translocation of Micronutrients in Plants," in *Micronutrients in Agriculture,* ed. R. C. Dinauer [Proceedings of a symposium, Muscle Shoals, Alabama, April 20-22, 1971. Madison, Wis.: Soil Science Society of America, 1972], pp. 202, 213).

9. K. L. Segraves, *Sons of God Return* (New York: Pyramid Books, 1975), p. 136.

10. Ibid.

11. Ibid.

12. Editors of Consumer Reports Books, *Health Quackery* (Mount Vernon, N.Y.: Consumers Union, 1980), pp. 218-37.

13. H. M. Morris, *Scientific Creationism,* General edition (El Cajon, Calif.: Master Books, 1974), p. 212.

14. Y. Heslop-Harrison, "Carnivorous Plants," in *Life at the Edge,* eds. J. L. Gould and C. G. Gould (New York: W. H. Freeman, 1978), pp. 92-106.

6

Creationism's Animal History

In the standard evolutionary view, the first organisms arose in the Precambrian, possibly 3.4 billion years ago, and the first animals evolved late in the same era (or early Cambrian) about 700 million years ago.[1] Biblical literalists believe that all basic "kinds" of life were created in three days, within one week, less than 10,000 years ago.

Framework of Biblical Animal History

Fundamentalist creationists believe that one authentic account of ancient history exists and that faithful interpreters and expositors of the Bible, particularly its first eleven chapters, are the only reliable custodians of that history.

Figure 4 depicts an abbreviated creationist history of animals from the alleged creation to the present. The branching patterns in the figure are similar to those given by Whitcomb.[2] For animal classification, creationists employ a one-size-fits-all, generic term called "kind" (sometimes expanded to "basic kind" or "created kind"), which can mean virtually anything from "species" to "order."

According to creationists, the deity who created the universe also pronounced curses on all things, including humans and animals. After the curse, populations increased throughout two biblical periods: 1,656 years before the flood and about 4,350 years after the flood.

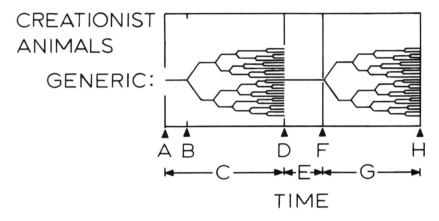

Fig. 4. Basic doctrines in creationist animal history. The branching diagram of generic animal kind, biblical chronology, and events are discussed in the text. Symbols: A. Biblical creation. B. Curse in Eden. C. Preflood period. D. Universal flooding. E. Flood period. F. End of flood. G. Postflood period. H. Present.

The universal flood allegedly continued on the earth for 371 days, drowning or freezing all animals left outside a legendary ark. Creationists contend that since creation, except for the flood year, population increase and diversification within animal kinds have continued to the present—without the evolution of new kinds.

The following are summaries of creationist history discussed in later sections.

Snake-Fable Foundation. A snake fable is the base for much of creationism's animal history and purported zoological science. Humans received counsel from a serpent and fell into sin. Sin brought curses on living things and elements of the soil. Under the curse, animals suffered profound changes in body structure and instincts.

Changes in Animals. All animals originally were vegetarian. Because of human sin, animals acquired new body parts designed for predation and defense. Some vegetarians became meat eaters; others became bloodsuckers. Biblical birds included bats. More birds existed at creation than exist now.

Creationism's Animal Kinds. Creationists have proposed definitions for animal "kind" that cover species, genera, family, and even order. This convenient classification strategy allows them to accept all organic and structural changes that have occurred and to manipulate any selected animal within a broad taxonomic range. They also can declare that no new species ever evolved, because every animal had to remain within the narrow limits of its kind.

Genesis Flood and Frozen Animals. Creationists believe that global ice-free conditions prevailed before the flood. The falling canopy water turned to ice and snow and glaciers. Water and the sediments in which frozen animal bodies are found were somehow cut off from warm oceans, they speculate, and quick-frozen at many northern polar sites, where they remain in permafrost.

Fundamentalist Faith: Snake-Fable Facts

Few ideas have had greater influence on Western culture than the idea of original sin. In Christian theology, an alleged sinful choice by Adam caused every person of the human race to inherit a depraved nature and experience evil effects of that choice.

The beginning of the story is the fable told in Genesis 3. Briefly, a talking serpent counseled and deceived the first woman, who then ate and tempted her husband to eat a forbidden fruit. This allegedly triggered curses from an angry God on all the creation.

To discuss alleged effects of the fall, I take the statements of Genesis 3:14–19 in reverse order. The man was sentenced to sweat and toil all his life to wrest a living from the ground, which would also yield thorns and thistles. And the woman would experience greatly multiplied pain in childbearing. Whitcomb and Morris accepted Genesis 3:16 as definite evidence that a "change took place in Eve's body."[3] They elaborated on this notion by asserting that ". . . the very structure of her body was now altered by God . . ." so that childbirth would be severely painful. These authors appeal to faith; they never identify "the very structure" nor state how it was altered.

Would snake-fable facts and related notions about the altered female body become part of human anatomy courses if creationists were given teaching authority in public schools?

Snakes: Creationism's Onetime Quadrupeds

The bodies and behavior of snakes, in the creationist view, reflect the curses of God in a most visible manner. Whitcomb and Morris assumed that snakes had four legs before Adam sinned and that the curse of having to crawl on their bellies meant that their legs were removed.[4] Without giving any evidence, these authors concluded, "Surely to be deprived of limbs involved far greater structural transformations in this creature than would have been involved in changing herbivores to carnivores. . . ."[5] Concerning feeding habits, they emphasized that every living thing (Genesis 1:30) had a diet of green herbs, thus there were no carnivores before Adam's sin.[6] Therefore, snakes originally were quadruped vegetarians.

Creationists emphasize repeatedly that no matter how radical the structural and organic changes involved, animals never became new species; they maintained their identity of "kind."[7]

To some observers, the feeding habits of snakes are astonishing. Snakes have fangs and only small needlelike teeth not capable of chewing; they swallow their prey whole. We ask, therefore, of what value in a strictly vegetarian diet are teeth without food-grinding ability? Creationists might respond that snakes originally had proper chewing molars, but their mouths were totally restructured when Adam sinned.

What are the ramifications of such imagined changes? In addition to the molar extraction and retrofitting process, snakes also received loosely hinged jaws that would allow them to swallow prey several times larger than their own body diameter. Certain poisonous vipers were fitted with ducts and glands and hollow fangs for venom injection. They were also given infrared sensing ability to detect heat from nearby prey. The changes weren't simple.

One would think that a serpent, an alleged herbivore, radically restructured and transformed into a deadly carnivore—one that also lost its legs—should be called a new species. But the creationists are adamant: No new species!

Creationism's no-new-species conjecture would, of course, extend to anteaters, lice, leeches, and thousands of other animals that creationists must assume were transformed to flesh-eating or blood-sucking types. To support their views, creationists need to invent new models of animal behavior and ecology for many thousands of parasitic species. This obviously includes everything that controls an animal's acquisition of food and feeding activity. So far, creationists have not intimated that anteaters and leeches had vegetarian diets before the alleged entry of sin, nor that the larvae of parasitic wasps were vegetarians before the presumed switch to living flesh of paralyzed spiders or other prey.

Eden's Curse of Fang and Claw

Responding to a theologian who questioned the utility of cat's teeth and claws for a strictly vegetarian diet, Whitcomb and Morris asserted that the questioner had completely missed the point.[8] "The point," they explained, "is that such specialized structures appeared for the first time after the Edenic curse."

Many creationists who interpret the Bible literally apparently believe that no animal ever killed or ate flesh before the biblical fall and resulting curse. Such alleged events provide creationists with a model for virtually endless conjecture about profound heritable changes in the animal world, changes that purportedly came abruptly not more than 10,000 years ago.

Creationists need to give account for more than cat's teeth and claws. Other animals have mandibles, stingers, horns, quills, spines, bony shells, and chemical weapons. The strong talons and keen eyesight of raptors would certainly not have been necessary for spotting and capturing vegetation. Nor would animal instincts for stalking prey have been useful in a vegetarian existence before the fall. (I must note here that many marine and fresh water animals—about which creationists are remarkably silent—also have similar physical makeup and instincts for taking prey.) Do creationists believe that drastic structural change and new instincts and capabilities all came on that single day of alleged human transgression?

The skeletal forms of extinct animals, particularly the dinosaurs, have very prominent and diverse features of anatomy.[9] The *Monoclonius, Pentaceratops,* and other dinosaurs had one to five massive head-horns and large bony hoods extending from the skull over the neck. Bony plates covered the body, and rows of spines were on the tail and sides of *Hylaeosaurus. Stegosaurus* carried pairs of upright spikes on the end of the tail. *Ankylosaurus* had heavy plates over the body and a massive bony club on its tail. *Allosaurus, Tyrannosaurus,* and other dinosaurs had very large jaws and recurved teeth. *Deinonychus* literally possessed the "terrible claw."

Whitcomb and Morris asserted that cat's teeth and claws were the result of a curse. Would these authors assert that the prominent structures of the dinosaurs named above also appeared after the Edenic curse?

John D. Morris, administrative vice president of ICR, instructed readers on alleged dinosaur history. He admitted that the Bible doesn't tell when dinosaurs began to eat meat, but he believed that it does give a clue: the sinful rebellion of Adam and Eve. Morris claimed that in pronouncing a curse of death,

> . . . evidently He [God] actually changed the genetic makeup of each "kind" so that all their descendants would forever be different. . . . Perhaps at this time dinosaurs and other animals acquired or began to acquire body parts designed for aggression or protection.[10]

"At any rate," he asserted, "Biblical history has an explanation for dinosaurs, their creation, life style, and extinction." Finally, he encouraged parents to "use" dinosaurs "to teach Biblical truth." The verb "use" is a conspicuous choice. Morris's call for parents to use dinosaurs is actually a call to abuse them. No explicit history of life styles and extinctions of dinosaurs can be found in the Bible.

Chapter 11 further analyzes creationist views on dinosaurs, including John Morris's alleged Paluxy dinosaur and mantrack history. The final section of chapter 11 gives a review of a children's book by Paul Taylor[11] in which he depicts dinosaurs as harmless, plant-eating pets and servants of people.

Creationism's Birds, Bats, Beetles, and Cats

Religious fundamentalists, particularly the current militant creationists, are striving to instill Bible-based traditions in America's schools. What might the curriculum and lessons contain if dedicated biblical literalists had authority to prepare classroom studies on birds? The following are possibilities.

The Command Formula: "Let Birds Fly." Among the most puerile of the biblical teachings about birds is that they fly above the earth. Creationists undoubtedly would emphasize the alleged command at the creation: "And God said . . . let birds fly above the earth across the firmament of the heavens." This, of course, leaves out the flightless ratite birds such as the ostrich, penguin, cassowary, kiwi, emu, rhea, and the extinct moa. With the exception of the ostrich, did biblical authors know about such birds?

Unclean Birds (Don't Forget the Bats). Biblical statements about certain "unclean" birds (Leviticus 11:13–19) could provide further lesson material for teachers who consider Bible testimony infallible. The passage names unclean birds, including the bat. Would creationist teachers dare contradict the Bible by telling students that the bat is a mammal and definitely not a bird?

Although bats make up a very diverse order containing more than 850 species, Gish identifies bats as a single basic kind.[12]

One Pair or Seven Pairs? Creationists could be expected to teach that Noah took two of each bird species into the ark (Genesis 6:19–20, 7:14–15). Sandwiched between the above two passages, however, is an apparent revision and a contradictory command having to do with alleged clean and unclean animals. Noah was instructed to take seven pairs of clean and one pair of unclean animals but, specifically, "seven pairs of birds" (Genesis 7:2-3). Which version do creationists accept? One pair of birds or seven pairs? Why the contradictory command?[13]

How Many Birds? A reference on the world's birds lists 9,016 species.[14] How many birds have creationists recognized? How many at the biblical beginning? How many aboard the biblical ark? How many at present?

Obviously, 9,000 branching lines for existing bird species cannot be diagrammed on figure 4. Leading creationists have chosen to show on such diagrams only a few classes of animals, e.g., man, horse, and dog[15] and man, gorilla, dog, and cat.[16] Biblical literalists, of course, hope to capture evolutionary terms, e.g., microevolution, and use them for their own purposes. Simple branching from a single line (see fig. 4) might convince those untrained in science that microevolution could account for great divergence from parent kinds—without producing new kinds. But how can one understand Whitcomb's pronouncement that many thousands of distinct bird *species* existed as living kinds at creation?

This is Whitcomb's statement:

> At present there are almost 8,600 distinct species of birds in existence, but there were more in the beginning than there are now. . . . Living kinds can (and many already have) become extinct; but none can ever evolve.[17]

If more bird species existed in the beginning, Whitcomb should specify the time of origin and the number of birds he can offer to support his statement. Many problems are left unsolved. No one has identified specific fossils of those alleged bird species in excess of present numbers. Nor has anyone produced evidence that bird species could never evolve.

Beetles and Cats. The two branches of the "generic" diagram (see fig. 4), one before the flood and one after, are identical and thus conform generally to those published by Whitcomb.[18] Obviously, 350,000 beetle species cannot be represented on Figure 4. We therefore ask: If Whitcomb can assert that more bird species existed "at the beginning" than exist now, would he also have people believe that original beetle species exceeded the number existing now? One might wonder about the 292 genera and approximately 8,800 known species of ants.[19] Were there more ants at the alleged beginning?

How can anyone make sense of Whitcomb's statements and accompanying diagrams?[20] Obviously, he wants everyone to believe that each kind (for example, cat) has a great potential for diversification—but never becomes another kind.

Guggisburg described members of the most highly developed terrestrial carnivores, the cat family, *Felidae*.[21] He discussed thirty-seven species (in sixteen genera), including lion, tiger, leopard, cheetah, cougar, lynx, and bobcat. He also named four genera of sabertooth tigers. Apparently, creationists believe that all cat species constitute a single cat kind.

Whitcomb's diagram, labeled "CATS,"[22] started as one preflood ancestral line and diversified to twenty-six descendant lines. After the Genesis flood a single line (one biblical feline pair allegedly from Noah's ark?) again diversified to twenty-six descendant lines. Whitcomb had emphasized that animal kinds have "amazing potential for variation."[23] Does Whitcomb believe that all descendants of a kind—such as his "CATS"—were interfertile and could interbreed and produce viable offspring? Does he believe that in a mere 4,500 years (after the alleged flood) two ordinary house cats or any mixed cat pair (lion and tiger, or leopard and house cat) could have reproduced all of Guggisburg's thirty-seven cat species known today?

In summary, the creationist model involves land-animal kinds rapidly spreading and diversifying in a period less than 2,000 years (see fig. 4, points B to D), after which selected animals passed through the biblical flood. (Creationists are silent about any rescue and preservation of marine life.) Point F represents new beginnings. Each selected animal kind increased greatly and diversified to the present (almost 4,500 years biblical time)—always within the bounds of its kind. Thus, present animal species allegedly retain a genetic identity reaching back to point F and, finally, to the alleged curse that brought profound genetic and structural changes in parent kinds and all their descendants.

Many scientists are involved in studies of genetic processes as well as the momentum and mode of evolutionary change in animals. I am not aware of any scientist who doubts that genetic mutations and selective pressures of environmental forces have important effects on the genetic constitution, the fertility, and the phenotypic traits of animal species.

Kitcher concludes that the interfertility criterion—which creationists frequently invoke in their discussions—falls short as a means of identifying basic kinds:

> There are species of *Drosophila*, for example, whose chromosomes are clearly derived from a common ancestor, but that divide up the chromosomal matter differently. Most of these species are unable to produce fertile hybrids. Some are unable to produce hybrids at all.[24]

Philosophic Difficulties with "Basic Kinds"

A problem scientists cannot ignore is that creationists have no coherent view of basic animal kinds. Opinions vary over a wide range, and some are clearly contradictory.

I summarize from Kitcher[25] the following creationist definitions and explanations for animal kind: (1) Kinds refer to species. (2) Kind may refer to genus. (3) Kind implies interfertility of similar organisms. (4) In the absence of fertility among similar animals, morphology can disclose sameness of kind.

Creationist opinions about kinds, however, do not stop with species and genus. Some place kinds in taxa of higher categories. Wayne Friar, a zoologist and creationist, was critical of evolution and offered several opinions about animal kinds during the Arkansas Trial.[26] Friar guessed that the originally created kinds numbered around ten thousand. His thinking and that of his fellow scientists about created kinds was inclined more toward the concept of family, but could include order in some cases. From his long experience in studying turtles, however, he could not say that all turtles are within the same created kind.

To the statements above, we may add another from a leading creationist, Duane Gish. He stated, "A basic animal or plant kind would include all animals or plants which were truly derived from a single stock."[27] Kitcher has criticized Gish's vague definition about descendants from an original kind, because ". . . the whole point of dispute concerns which organisms descend from the same original stock."[28] For those who wish to consider refutations of creationist views from the perspective of systematics and paleontology, I recommend the article by Cracraft[29] and books by Eldredge[30] and Newell.[31]

Cracraft examined the widespread confusion creationists reveal in their use of the created-kind concept. He described how Gish contradicted the basic assumption on which the concept rests.[32]

Gish presented several lists of animal kinds.[33] He asserted that, "Among the vertebrates, the fishes, amphibians, reptiles, birds, and mammals are obviously different basic kinds." He named twelve mammals, including, ". . . dogs, cats, lemurs, monkeys, apes, and man. . . ." These also he recognized as "different basic kinds." Finally, he advised, "Among the apes, the gibbons, orangutans, chimpanzees, and gorillas would each be included in a different basic kind."

Gish's arrangement of animal groups in a related series was a mistake; it turned upside down the biblical dogma of permanently separate, created kinds. The vertical order he established between his suggested groups exhibits an organic succession—the evolutionary descent with modification—exactly what his opponents have been saying.

Cracraft expressed the issue succinctly: "Independently 'created kinds' cannot be hierarchically arranged, one within the other."[34] Cracraft clearly illustrated in his Figure 1 the hierarchy in Gish's organized kinds.[35] He concluded, ". . . it is logically not possible for one 'created kind' to be a member of another 'created kind.' "

Kitcher imagined an unguarded moment when Gish placed the "gibbons. . . . gorillas" kinds within another presumed kind, the apes.[36] He responds in a striking metaphor, "It seems that Creationists will strain at finches and fruit flies only to swallow the apes."

No creationist or scientist or any other person has understood and described with certainty what creationists mean when they talk about kinds. The vagaries and contradictions inherent in the notion of kind render the term useless for scientific discussion. Gallant considered the term "basic kind" to be ". . . virtually meaningless and never . . . found in scientific literature outside the creationist community."[37] Godfrey stated that, "In fact, the definition of 'created kind' is one of the most confused issues in the creationist literature. . . ."[38]

Creationist Flood and Fabricated Quick-Freeze History

People have enormous interest in mammoth, mastodon, and other animal graveyards in polar regions. Aware of such interest, creationists hope to establish that millions of animals were frozen and buried by the Genesis flood about 4,500 years ago. Their strategy has been to make the animal freezing deaths fit their catastrophic flood model, just as they also have tried to force long evolutionary sequences of the geologic column, and corresponding fossils, into a neat slot of a single flood year.

In discussing Arctic soils (under "post-deluge geologic activity"), Whitcomb and Morris stated that regions of permanently frozen soils, called *permafrost*, were formed "at some intermediate or late stage in the Deluge period. . . ."[39] And buried in that permafrost, they noted, are many mammal fossils.

Creationist-watchers would like to know what they mean by "intermediate stage." Does their late stage mean that alleged "post-deluge" activity formed the permafrost? Apparently Whitcomb and Morris deliberately chose to be vague about time. They tried to explain that the vapor canopy had profound effects on climate when it collapsed and fell to earth.[40] But the turbulence and warmth of the polar seas, they assumed, prevented freezing for an unknown, but considerable, time. In their view freezing did not occur early, but they don't know when.

Since these authors could not decide when the freezing took place, perhaps they determined where and something about the watery medium being frozen. They stated that,

> Undoubtedly, the first water actually to freeze would have been that mixed in with the sediments being deposited in these regions, cut off, as it were, from the warmer temperatures and turbulent agitation of the free water in the open ocean.[41]

They repeat the same emphasis about sites isolated from warm seas.[42]

Whitcomb and Morris seemed very confident about sediments

and trapped water being cut off from warm ocean water, but they made no effort to identify isolated regions. Neither have they identified elevated terrain or any other barrier that might isolate freezing sediments. Rather, they discussed features of lowlands: frozen mucks, deposited sediments, silts, and alluviums.[43] And they cited authorities who mention Arctic Islands, silty alluvium in Alaska, and rivers flowing into the Arctic Ocean. Nowhere do they identify barriers to ocean water incursion.

Ironically, Whitcomb helped to dispel the notion of sites being "cut off" from warm oceans. He noted that "An estimated 5,000,000 mammoths, whose remains are buried all along the coastline of northern Siberia and Alaska, were frozen and buried not many years ago."[44] Animal remains all along coastlines do not support the assumption that animals were isolated from oceans.

Do creationists ever doubt the surmised role of their collapsing vapor blanket in the quick-freeze death and burial of animals? After discussing Siberian mammoth beds and other remains in Arctic Islands, Whitcomb and Morris offered this oblique admission, "It may not be quite clear as yet whether these deposits were made directly during the Deluge period or soon after, or both. . . ."[45] What forced these authors to make their vague and ambiguous statement about animal deaths and burial? If they could have spoken with candor, their it-may-not-be-quite-clear apology could have been a forthright statement of fact: the whole quick-freeze scenario is not clear, and they don't know when the Arctic animals died.

Creationists are left with absurd choices for Arctic animal deposits in their flood period. Deposits (burials) made during a universal flood would have been overwhelmed with water, not ice and snow. Deposits made after the alleged flood year would mean that bodies had floated around for months before burial, and the flesh would not have shown the integrity of quick-frozen tissue. And if the burial of Whitcomb's "estimated 5,000,000 mammoths" came in a late phase of the "Deluge period," how late was it? Late enough for two mammoths to increase to 5,000,000? Obviously, creationists are trying very hard to push the mammoth story into their flood story.

In more recent writings, several creationists returned to the early-phase notion of flood activity in the deaths of Arctic animals. John Whitcomb, in 1961, could admit confusion about the time the animals were frozen. But later he apparently had changed his mind about when the fast-freeze occurred.[46] He quoted a report on a frozen mammoth, with food undigested, and proposed that quick freezing such animals would have required, ". . . temperatures of 150 degrees below zero . . . to descend upon them instantly." He then concluded that in the "early stages" of the canopy's collapse such fast-freeze conditions "could very well have existed" in Arctic regions.[47]

For Whitcomb, it was a freezing process occurring instantly in an early flood stage. Was Whitcomb referring to floodwater that had, by creationist conjecture, risen 175 feet on the first day? Or 500 feet in three days? We should note that if the maximum height of floodwater occurred in forty days, as Whitcomb and Morris claimed,[48] water had to rise an average 175 feet per day to cover creationism's 7,000-foot mountains.

In 1986 Whitcomb was still writing about frozen canopy water and sudden freezing deaths. He added new speculations. His canopy vapor turned to snow and ice and then to glaciers in the flood's early weeks. He says,

> The collapse of this vapor canopy during the first weeks of the flood (Genesis 7:11–12) took the form of snow and ice in the higher latitudes, causing huge glaciers, the sudden freezing of mammoths and other creatures. . . .[49]

During the first weeks? If we select two weeks, that would mean, in Whitcomb's view, that the equivalent of a 2,500-foot (760 meter) depth of liquid water (covering earth at lower latitudes) would have occurred in the Arctic and Antarctic as massive ice packs and glaciers. What icy forms existed when the imagined flood engulfed their 7,000-foot mountains in forty days? Was their flood-caused permafrost still in the ground?[50]

It is ironic that creationists have named environmental reservoirs where the proposed heat content could invalidate their simplistic freezing scenario. Some of the reservoirs listed below are described as extremely hot; others are warm: (1) Massive vapor canopy at a possible 3,000°F. (2) Exploding "suboceanic and subterranean" reservoirs of "magma and steam." (3) Lower atmosphere (heavier gases) allegedly warm everywhere. (4) Universally warm ocean. (5) Land masses dominated everywhere by a seasonless warm climate.

Will creationists ever try to describe all of water's interactive and heat-loss phases as it descends from superheated vapor to an icy state 150°F below zero? How could superhot vapor lose heat to form water droplets (at 706°F, the transition point)? How could water cool farther to become ice while passing into reservoirs of heated air and vented heat from universally exploding caverns? How could ice remain frozen despite exposure to radiated heat of continents and warm, turbulent oceans? And how could ice descend finally to a supercool temperature that would freeze large animals instantly?

Creationists talk much about "thermodynamics" but very little about dynamics of heat flow in their primitive world. It is crucial that they sort out the factors and physical conditions that would allow them to analyze critically their quick-freeze equation.

More Fast-Freeze Fantasy

ICR associate professor of meteorology Larry Vardiman believed that a great vapor canopy once enveloped the earth and caused a universally warm climate.[51] Because of that climate, he assumed, many animals lived near the poles but were frozen in place when the canopy collapsed and the flood came. Vardiman then walks into his own rhetorical trap: "Animals caught in the flood, cold, and wind would be frozen rapidly along with the sediment from the flood."

Vardiman's assumptions do not make sense. Even those un-tutored in hydrology and flood science can understand that to be caught in a flood is to be caught in liquid water. And to speak of "sediment from the flood" is to recognize processes of erosion by liquid water and the transport of the suspended materials by the same liquid phase of water. Contradictory events cannot occur at the same time. Animals being overwhelmed by turbulent, rising water were not also being quick-frozen in place.

Vardiman expanded his quick-freeze story. He referred to fresh vegetation in stomachs of frozen mammoths as proof of a sudden temperature drop, and, noting the preserved state of mammoth flesh, he concluded that the drop was both extreme and perma-nent.[52] Again, readers must ask for reasonable explanations for a fast-freeze process occurring in turbulent water. Moreover, how could freezing temperatures be permanent concurrent with flood-water rising continuously and finally covering all the high mountains?

After unconvincing arguments, Vardiman came to this gratuitous conclusion, "Such a scenario matches the predictions of the vapor canopy very well."[53] Vardiman's scenario does not rest on credible evidence. Animal flesh does not quick-freeze in liquid water. As for his scenario being predicted from the canopy model, no one has verified the canopy's existence.

Animal Studies: Challenges for Educators

Religious fundamentalists who interpret the Bible literally present serious challenges to the scientific community and those involved in public education. Of particular concern are creationists whose primary goal for animal studies is that Bible-based myths be taught in all schools. This would include several of the subjects raised in this chapter: (1) Origin of limited created kinds of vegetarian animals less than 10,000 years ago. (2) Sinful disobedience and curse-induced genetic and structural transformations in animals. (3) Transformations of animals from herbivory to carnivory and parasitism. (4) Scheduled extinctions of animals by drowning and freezing in a recent universal flood. (5) Repopulation of earth by selected descendants of original animal kinds.

Some points made in the above statements come directly from Genesis 1–11. Other points arise from creationist expansions and speculation (e.g., animal deep-freeze stories). Numerous terms in the five statements above embody creationist notions about origins, taxonomy, anatomy, physiology, genetics, reproduction, ecology, paleontology, and related disciplines. The intention of political activists among the religious fundamentalists is to force all knowledge and teaching in biology into agreement with their interpretations of the Bible.

Because creationism's animal history derives from mythical Bible stories, many scientists and educators have dismissed it as totally mythical in origin and unreliable. Their history incorporates elements of mystery. It requires supernatural acts and curses of a deity. It displays the anecdotal character and superstition of folklore. It actually elevates a snake fable to the level of absolute fact.

The health of America's biological teaching programs will depend on continued scientific research and on crucial decisions made by the nation's educational communities, including parents, teachers, principals, school boards, and librarians. To the role of the local community we must add the powerful influence of county, state, and national organizations and individuals. These include educational associations, boards of education, superintendents, secretaries, legislators, judges, and justices.

What weight will that total community give to Bible stories? Will educators reject the biblical snake fable, flood myth, and related creationist lore as an unworthy base on which to build school curricula in biological science?

Notes

1. G. G. Simpson, *Fossils and the History of Life* (New York: Scientific American Books, 1983), p. 59; A. N. Strahler, *Science and Earth History: The Evolution/Creation Controversy* (Buffalo, N.Y.: Prometheus Books, 1987), pp. 312–13.

2. J. C. Whitcomb, *The Early Earth* (Grand Rapids, Mich.: Baker Book House, 1986), p. 95.

3. J. C. Whitcomb and H. M. Morris, *The Genesis Flood* (Phillipsburg, N.J.: Presbyterian and Reformed Publishing Co., 1961), p. 466.

4. Ibid., p. 464.

5. Ibid., p. 465.

6. Ibid., p. 461.

7. Ibid., pp. 464, 466.

8. Ibid., p. 465n.

9. H. R. Sattler, *Dinosaurs of North America* (New York: Lothrop, Lee and Shepard Books, 1981); D. F. Glut, *The New Dinosaur Dictionary* (Secaucus, N.J.: Citadel Press, 1982); D. Cohen, *Dinosaurs* (New York: Doubleday, 1987).

10. J. D. Morris, "How Do the Dinosaurs Fit In?" in *Back to Genesis* (El Cajon, Calif.: Institute for Creation Research, May 1989), p. d.

11. P. S. Taylor, *The Great Dinosaur Mystery and the Bible* (El Cajon, Calif.: Master Books, 1987).

12. D. T. Gish, *Evolution? The Fossils Say No!* Public school edition (San Diego, Calif.: Creation-Life Publishers, 1978), p. 47.

13. Genesis 7:2–5 apparently is the work of special-interest revisionists, the devotees of ritual cleansing and sacrifice. Such persons would have realized that to sacrifice either member of a single pair would mean extinction of a species. Editing the narrative to include more than single pairs of clean species agrees with the ritual need to offer only clean animals to their God. The editing also is consonant with the alleged later offerings of the clean animals and birds as burnt sacrifices (Genesis 8:20, 21).

14. J. J. Morony, Jr., W. J. Bock, and J. Farrand, Jr., "Reference List of the Birds of the World," *American Museum of Natural History* (New York: Department of Ornithology, 1975).

15. Whitcomb and Morris, *The Genesis Flood,* p. 67.

16. Whitcomb, *The Early Earth,* p. 95.

17. Ibid., p. 112.

18. Ibid., p. 95.

19. B. Hölldobler and E. O. Wilson, *The Ants* (Cambridge, Mass.: Belknap Press of Harvard University Press, 1990), p. 4.

20. Whitcomb, *The Early Earth,* pp. 94–95.

21. C. A. W. Guggisburg, *Wild Cats of the World* (New York: Taplinger Publishing Co., 1975), pp. 23–289.

22. Whitcomb, *The Early Earth,* p. 95.

23. Ibid., p. 94.

24. P. Kitcher, *Abusing Science* (Cambridge, Mass.:MIT Press, 1982), p. 153.

25. Ibid., pp. 151–55.

26. R. A. Gallant, "To Hell with Evolution," in *Science and Creationism,* ed. A. Montagu (New York: Oxford University Press, 1984), p. 296.

27. Gish, *Evolution? The Fossils Say No!* p. 34.

28. Kitcher, *Abusing Science,* p. 152.

29. J. Cracraft, "Systematics, Comparative Biology, and the Case against Creationism," in *Scientists Confront Creationism,* ed. L .R. Godfrey (New York: W. W. Norton, 1983), pp. 189–205.

30. N. Eldredge, *The Monkey Business: A Scientist Looks at Creationism* (New York: Washington Square Press, 1982).

31. N. D. Newell, *Creation and Evolution: Myth or Reality?* (New York: Columbia University Press, 1982).

32. J. Cracraft, "The Significance of the Data of Systematics and Paleontology for the Evolution-Creationism Controversy," in *Evolutionists Confront Creationists,* eds. F. Awbrey and W. M. Thwaites (Proceedings of the 63rd Meeting, San Francisco: Pacific Div., *AAAS,* vol. 1, part 3, April 30, 1984), pp. 193–94.

33. Gish, *Evolution? The Fossils Say No!* p. 35.

34. Cracraft, "The Significance of the Data," p. 194.

35. Ibid.

36. Kitcher, *Abusing Science,* p. 154.

37. Gallant, "To Hell with Evolution," p. 295.

38. L. R. Godfrey, "Scientific Creationism: The Art of Distortion," in *Science and Creationism,* ed. A. Montagu (New York: Oxford University Press, 1984), p. 174.

39. Whitcomb and Morris, *The Genesis Flood,* p. 288.

40. Ibid.

41. Ibid.

42. Ibid., p. 290n.

43. Ibid., pp. 288–90.

44. J. C. Whitcomb, *The World That Perished* (Grand Rapids, Mich.: Baker Book House, 1973), p. 77.

45. Whitcomb and Morris, *The Genesis Flood*, p. 289–90.

46. Whitcomb, *The World That Perished*, p. 77.

47. Ibid., p. 80.

48. Whitcomb and Morris, *The Genesis Flood*, pp. 4, 8.

49. Whitcomb, *The Early Earth*, p. 144.

50. Whitcomb and Morris, *The Genesis Flood*, p. 290n.

51. L. Vardiman, "The Sky Has Fallen," *Impact*, no. 128 (El Cajon, Calif.: Institute for Creation Research, Feb. 1984): i–iv.

52. Ibid., p. 5.

53. Ibid.

7

Animal Migrations and Ark Experiences

Creationism's Big Boat

The flood narrative (Genesis 6:15) specifies dimensions for Noah's ark in cubits: 300 long, 50 wide, and 30 high. The accepted value for the cubit is 18 inches, which gives the ark an alleged length of 450 feet. Wonder, even intimidation, can be the expected response of some impressionable persons who hear such large numbers. Dwarfed in size, they are overwhelmed and unable to think critically about the assumed seaworthiness of the mythical boat. I have taken the following information and quotes about wooden ships from Robert Moore.[1]

1. The length limit for wooden boats is about 300 feet.

2. Beyond 300 feet deformation becomes excessive. and maintaining a watertight hull becomes increasingly difficult.

3. Long wooden boats are liable to sag and hog (sagging at the ends and bowing up amidship), which is "the major reason why the naval industry turned to iron and steel in the 1850s."

4. "The largest wooden ships ever built were the six-masted schooners, nine of which were launched between 1900 and 1909. These ships were so long that they required diagonal iron strapping for support; they 'snaked,' or visibly un-

dulated, as they passed through the waves, they leaked so badly that they had to be pumped constantly, and they were only used on short coastal hauls because they were unsafe in deep water."

5. The longest six-master, 329 feet, was the U.S.S. *Wyoming.*

The length of the biblical ark allegedly was more than 100 feet longer than the six-master, U.S.S. *Wyoming.*

Creationists do not admit that the ark could have twisted or have been deformed under the impact of huge waves. Henry Morris asserted, "The ark . . . was admirably designed to ride out the approaching storm in comparative comfort."[2] He continued, "Both hydrodynamic calculations and laboratory wave-tank model testing have demonstrated that the ark was so dimensioned as to be exceedingly stable in the violent waters of the flood."

The scientific community should challenge Morris to produce those wave-tank procedures and calculations that prove the exceeding stability of a 450-foot wooden boat. Further, who reviewed the data? And where were they published?

Morris made other questionable statements: "It [Noah's ark] was, once loaded, practically impossible to capsize, and would align itself in such a direction as to ride the waves most comfortably."[3] How, we ask, could a rudderless barge align itself to waves? What specific features of the design accounted for that ability? And what was the necessary direction of alignment to oncoming waves? Scientists, I suspect, will regard those statements by Morris only as fictions if he doesn't produce scientific evidence.

Creationist commentators have given considerable detail—not found in the Bible—concerning ocean depths and Noah's ark. Genesis 7:19-20 states that ocean water "prevailed above the mountains, covering them fifteen cubits deep." Whitcomb and Morris advised readers that to appreciate the true significance of the 15-cubit depth, one must understand that the ark sank 15 cubits into the water when fully loaded.[4]

This raises a serious problem: the ark could not have cleared the highest of the submerged mountains. Did the ancient writer comprehend that his ark might have run aground? Commentators Whitcomb and Morris were also concerned about the ominous and menacing character of great waves in the flood.[5] The "tsunamis," the most destructive of all waves, they believed, "must have been produced" in the flood.[6] And because of volcanic upheavals in the ocean, they concluded, "Great tidal waves undoubtedly were generated in prodigious numbers. . . ."[7]

Apparently, their own worries about perils of monstrous waves caused Whitcomb and Morris at the beginning of their book to take up the problem of the ark's draft.[8] Could they also have recognized the specious accuracy of the biblical 15-cubit claim? In any case, they revised the Bible's fifteen cubits to a depth of "*at least* fifteen cubits" (emphasis in original).

Not surprisingly, the biblical author needed some expert editing of his story. Nor is it a surprise that modern creationist editors must get more water between those mythical, barely submerged mountains and the big boat.

Five Hundred Twenty-Two Boxcars

Creationists seem confident that describing Noah's ark in terms of hundreds of railroad boxcar volumes will prove to everyone that the big boat had more than adequate space for many thousands of animals.

The staff and consultants of the ICR represented the carrying capacity of Noah's ark as being equal to "522 standard railroad stock cars." They concluded, "This is more than twice as large as necessary to accommodate two of every species of known land animal that ever lived."[9]

Creationists manipulate boxcars in a number of ways. Whitcomb pictured the ark as an enormous "flat-bottomed, square-sided barge" in which 522 boxcars could be stacked.[10] But arranging boxcars in a very long chain can also be very impressive—indeed may overwhelm credulous minds. For example, Whitcomb and Morris imagined, ". . . eight trains with sixty-five such cars in each!"[11]

By distributing 240 vertebrate animals in a two-decked stock-car (with each animal averaging the size of a sheep), these authors ensconced 35,000 animals in two, 73-car trains.[12] By that simple computation, they apparently believed that they silenced their opponents and disposed of objections "once and for all."

The same authors also divided boxcars into many small cubes, each two inches on a side.[13] They then assigned the space (eight cubic inches) to individual "insects, worms, and similar small creatures." By their calculation, twenty-one boxcars would have housed more than a million individuals. These authors calculated the standard boxcar volume as 2,670 cubic feet,[14] but then made a huge mistake. Since one cubic foot contains 216 cubes, a boxcar would have provided 576,720 cubes. Thus their make-believe story needed less than two boxcars, not twenty-one.

Segraves also repeated the same 522-boxcar fantasy.[15] He was amazed by the ark's great size and the purportedly small need for space. He accepted the Whitcomb-Morris figure of 35,000 animals in 146 boxcars. And he unwittingly added the twenty-one boxcars for insects and creeping things, allowing "insects two inches of flying room." Segraves's assignments of space to passengers gave one floor to known animals and the second floor to Noah and family. The entire third floor he reserved for recreation and perhaps some unaccounted-for extinct species.

Although creationists believed that Noah's ark had twice the space actually needed, they speculated about animals surviving the flood without shelter on the ark. Whitcomb and Morris proposed, ". . . many of the 25,000 'species' of worms, as well as many of the insects, could have survived outside of the Ark."[16] They gave no specific examples.

Can these authors name any insect or worm that could have survived an alleged forty days of torrential rain and an unbroken expanse of salt water for most of a year? In speculating about survivors outside, these authors apparently decided that land-dwelling animals allegedly on Noah's ark were not the sole survivors of the flood. Why did these literalist Bible interpreters contradict Genesis 7:23?

Summarizing the boxcar fantasy, I propose that creationists have fabricated a huge red herring. They seem very assured in reciting the ark's imagined attributes: more than adequate space, stability and safety, maneuverability, and its perfection as a vehicle of transport.

One rhetorical move an audience could miss, however, is the subtle switch from the ark's strictly transport function to one of habitation. Consider this analogy. A farmer prepares to deliver chickens to market. He loads the birds into bare crates and stacks them in an enclosed livestock van for transport. But then he decides to travel around for a year while the crated birds, he assumes, have no need for care—other than the security of the transporting van in which they will hibernate.

The big-boat and boxcar fabrications have remarkable red-herring quality. They fix people's minds on great numbers of animals and cubes—not on cubicles (dwellings, places habitable). People are led to think that transportability means habitability. They are not the same.

Creationists who talk of cubes and space for more than a million small creatures do not mention doors or passageways that would give access to cages or any of the necessary routines for feeding and watering. Rather, the boxcar metaphor creates the image of a warehouse where packaged matter can be closely stacked. One boxcar or 522, it makes no difference; boxcars are not habitats.

Boat Passengers: Adults, Babies, or Eggs?

The creationists who have tried to describe the total operation of the biblical flood and rescue legend have had to imagine and add many details to a relatively short account (Genesis 6:13-8:29). This virtually endless project has raised many questions, such as: What animals came to Noah? What caused their migration? Where did the animals live? How far did they migrate? Were there limitations on size and development of animals? Leaders in the creationist movement have given ready—and also unverifiable—answers to all these questions.

Selected male and female animals of every kind, allegedly received "a migratory directional instinct . . . to flee from their native habitats to the place of safety."[17]

Creationists reject long migrations to Noah's ark from isolated continents. Whitcomb's catechism on "Kangaroos and Noah's Ark" could be taken as typical of rote instruction in schools where creationist teachers have access:

Question: How could kangaroos have traveled from Australia to Noah's Ark? Answer: They didn't. At least two each of all the kinds of air-breathing animals—including kangaroos—must have lived on the same continent where the Ark was built. . . .[18]

Whitcomb continues:

> *Question:* How did kangaroos reach Australia from Mount Ararat after the Flood? *Answer:* A great land bridge apparently connected Asia and Australia in the early post-Flood period.[19]

Whitcomb conjectured that water held in polar ice lowered sea level to create a land bridge.[20] Despite evidence for numerous glacial and interglacial periods over millions of years, the creationists hold to one short ice age occurring about 4,300 years ago.

Whitcomb and Morris also conjectured ". . . that marsupials could have reached Australia by migration waves from Asia, before that continent became separated from the mainland."[21] Evidence noted in chapter 4 showed that the Indo-Australian plate has been moving toward the Asian continent, not away from it. The typical ad hoc strategy of some creationists writers is to manipulate things (continental plates, oceans, climate, ice ages) according to the needs of their developing stories.

Were there limitations on the size of passengers? The Bible sets no limit on size. Creationists writers, therefore, have filled in details. They know from fossil evidence that some dinosaurs were too tall for the ark, which was alleged to be 30 cubits, or 45 feet. Because alleged vertical space was limited (three floors in 45 feet), some creationists have excluded very large animals and limited passage to the very young. Another imagined solution has been to deny passage altogether, for example, to dinosaurs, "for the very reason of their intended extinction."[22] The biblical narrative does not mention such a destiny—rather, just the opposite (Genesis 6:19, 7:15).

John Morris speculated about dinosaurs taken on the ark—not necessarily adults, but younger, smaller animals. He suggested, "It is also possible that he [Noah] took dinosaur eggs on the Ark, eliminating the need for so much food and space."[23] One could wish he had disclosed something about possible egg collections. Did egg-laying dinosaur mothers get the divine urge to bring eggs to Noah? And did Noah select pairs of eggs for the ark's hatchery, eggs that should become earth's future breeding pairs? Despite John Morris's sincere suggestion, most creationists would probably consider Noah's leisure and a lizard hatchery not a workable combination. One reason is that the young of most species require immediate and continuing nutritional care for growth and development. Further, the noted leader among creationists, John Whitcomb, had already disclosed that Noah was relieved of all care for the ark's animals.[24] His view and the opinion of other leading creationists was that the only feasible plan for survival was hibernation.

Another imagined ark experience is quite remarkable. It is actually a prohibition, as if a sign had been posted at the ark's door: NO PREGNANT FEMALES ALLOWED. Whitcomb emphasized that God overpowered natural instincts and bodily functions in getting animals to the ark.[25] Whitcomb then disclosed earth's first birth-control program. He stated emphatically: ". . . there could have been no multiplication of animals (not even the rabbits!) during the year of the Flood, for the Ark was built just large enough to carry two of each. . . ." (The exclamation and rabbit statement are Whitcomb's.) Apparently the ark was too small for more passengers. However, in his later book Whitcomb needed only one-third of the ark to hold all air-breathing animal kinds now alive and those extinct.[26] The same notion of excess room had been emphasized in his earlier book.[27] It is puzzling how the size of Noah's ark could have changed so much from book to book.

Creationists have many more questions to ponder than those about ark dimensions, or animal size, age, and migration. Housing of insects for more than a year presents unique problems that creationists do not discuss. Apparently, they have no interest in the intermediate larval and pupal forms of insects. And because animals allegedly "came" to Noah, creationists have assumed that all insect migrations were by adults.

This raises crucial questions about stages in life cycles, particularly in those where one form in a cycle (egg, larva, pupa) exists much longer than the adult. For example, some nymphs of the cicada (called seventeen-year locust) live more than 800 times longer than the adult, which ordinarily lives about a week. A week is barely long enough for mating and egglaying for a new brood. Many other females of insect species live only a few days or weeks in which they mate and lay eggs. Creationists could appeal, of course, to miracles. They could declare that all hormonal and other processes resulting in reproduction were abolished, along with aging and death. But in such appeals they would advance religion, not science.

Genesis 7 gives explicit statements about animals entering Noah's ark. That account plus creationist ideas about animal numbers create a scenario difficult to understand and accept. The 35,000 larger animals (which creationists Whitcomb and Morris allow on board) came to the ark and entered, we are told, "on the very same day" that the rain started and the flooding began (Genesis 7:11–14). This miraculous, one-day loading schedule gave each animal, on average, about two seconds. Movement was through one door, and the boarding pace obviously included that of snails, slugs, turtles, snakes, sloths, and other slow movers, which also must avoid being crushed. This would have called for many miracles.

Creationists also allow one million creeping, crawling, and flying insects on board. All of them (if given a twenty-four-hour day) had 0.08 second each, or nearly twelve insects must enter each second. (Another million miracles?) Each passenger then found assigned compartments (sometimes referred to by creationist writers as rooms, stalls, nests, cages, or cubes) and entered for a year-long hibernation. (More miracles.)

The above scenario rests on biblical and creationist statements. Stated numbers of animals entered the biblical ark on a single day. The animals allegedly were the distant ancestors of today's populations. Is this what creationists call "biological science"? Is this part of the biological history they hope to teach in all American schools and, indeed, in the whole world?

Boat Ride on Volcanic Mount Ararat

Perhaps the most bizarre idea introduced by creationists concerns animals on a boat ride, for most of a year, on a fast-growing volcanic mountain. The main elements of the story come from the most prolific of the creationist writers, Henry Morris.[28] He referred to numerous reported sightings that "indicate that the Ark is still preserved in the icecap near the summit of the gigantic mountain known even today as Mount Ararat. . . ." He explained further, "This mountain is an extinct volcanic cone, formed probably during the Flood itself, now towering . . . 17,000 feet at its crest." Although the ark's presence is not documented, Morris reported that "much evidence" exists that the ark is ". . . hidden most of the time beneath the mountain's icy covering." Does the ark come out when no one is looking? The premise, "hidden most of the time" is pure conjecture. No ark has been documented outside the mountain's glacial covering and certainly not under it.

Briefly, Morris's alleged facts are these: an ark landed on a volcanic mountain, and the mountain with the ark near its summit rose to 17,000 feet. How far did Mt. Ararat rise in the flood year—after the ark's alleged landing? Creationist answers to this question are contradictory.

One answer is that the mountain and the ark rose from about 3,300 feet to 17,000 feet. A diagram by Whitcomb and Morris[29] depicts Noah's ark at the top of Mt. Ararat. Notes on the figure specify that floodwater receded "at an average rate of 15 feet a day" and that 221 days were required for complete withdrawal. Receding at that rate, the sea level would have dropped 3,315 feet, and Mt. Ararat's height would have been about 3,300 feet when the ark landed. If we take all this as fact, then the legendary boat ride was a vertical 13,700 feet in 221 days or an average 62 feet per day. This estimate, of course, is not final.

Twelve years after publishing *The Genesis Flood,* Whitcomb asserted that before the flood no mountains "were more than 6,000 to 7,000 feet high."[30] If we accept their value of 7,000 feet, then the vertical ride for the animals and the boat was 10,000 feet in 221 days, or 45 feet per day.

The development of the fundamentalist flood story seems not to have been very difficult. Leading creationist advocates and writers followed simple rules and kept within biblical constraints of a flood year allegedly occurring about 4,350 years ago. Their main challenges in writing have been to manipulate massive quantities of earth and water and a large number of animals. Not only must this occur within the flood year, said to be 371 days, but writers have had to deal with a limited supply of water. Before the story could be developed, therefore, they had to lower mountains and raise ocean floors to have enough water to cover earth's highest peaks.

Nor has the challenge of getting animals to the ark been difficult. Merely writing easy answers apparently solved migration problems. The simple solution? The home territories of selected pairs of animals were assumed to have been on the same continent and near the ark's construction site. This arbitrarily eliminated great distances and numerous barriers to migration.

A final responsibility for creationist writers was to disclose that selected animals came to the ark by divinely imposed instincts. By the same divine power animals went into a year's hibernation.

The basic creationist method is to build assumption upon assumption continuously, with none of them requiring scientific support before the next one is offered. This is the unchanging plan of the builders. Their result is an inverted pyramid of confusion.

I am not aware of any creationists willing to face problems of energetics in the alleged hibernation of great numbers of diverse animals. The nightly torpor of hummingbirds and the aestivation (periodic inactivity) of some animals do not correspond meaningfully to the imposed year-long hibernation of creationism's ark population.

Nor have creationists answered other energy problems: how animals in torpor for a year could have remained strong and have spread abroad, and how they found nutrient sources for reproduction—despite destruction of food reserves by a flood and too little time for revegetation.

Nor have creationists explained how a wooden boat survived massive lava ejections and heat radiation from a fast-rising volcanic mountain. They have not explained how animals on the rim of an erupting volcano were immune to toxic gases and heat.

Discussion of hibernation on a fiery mountain is, of course, meaningless for animals in a flammable boat. Creationists might invoke miracles to render the wood immune to flames; then somehow the boat must avoid burial under volcanic matter. Did it float on volcanic cinders and lava?

The fabricated answers creationists give to questions of mountain building and animal survival are an integral part of their geological and biological history. Educators can hope that such fabrications will never be certified for use in public schools. It is tyranny enough over innocent minds that such ideas are seriously taught in any institution.

Notes

1. R. A. Moore, "The Impossible Voyage of Noah's Ark," Creation/Evolution 11 (1983): 4–5.

2. H. A. Morris, *The Beginning of the World* (Denver, Colo.: Accent Books, 1977), p. 99.

3. Ibid.

4. J. C. Whitcomb and H. M. Morris, *The Genesis Flood* (Phillipsburg, N.J.: Presbyterian and Reformed Publishing Co., 1961), p. 2.

5. Ibid., pp. 261–64.

6. Ibid., p. 264.

7. Ibid., p. 261.

8. Ibid., p. 2.

9. H. M. Morris, *Scientific Creationism*, General edition (El Cajon, Calif.: Master Books, 1974), p. 253.

10. J. C. Whitcomb, *The Early Earth* (Grand Rapids, Mich.: Baker Book House, 1986), p. 98.

11. Whitcomb and Morris, *The Genesis Flood*, p. 68.

12. Ibid., p. 69.

13. Ibid., p. 69n.

14. Ibid., p. 68n.

15. K. L. Segraves, *Sons of God Return* (New York: Pyramid Books, 1975), p. 133.

16. Whitcomb and Morris, *The Genesis Flood*, p. 69.

17. Ibid., p. 74.

18. J. C. Whitcomb, *The World That Perished* (Grand Rapids, Mich.: Baker Book House, 1973), p. 25.

19. Ibid.

20. Ibid.

21. Whitcomb and Morris, *The Genesis Flood*, p. 87.

22. Ibid., p. 69n.

23. J. D. Morris, *Tracking Those Incredible Dinosaurs and the People Who Knew Them* (San Diego, Calif.: CLP Publishers, 1980), p. 66.

24. Whitcomb, *The World That Perished*, p. 32.

25. Ibid.

26. Whitcomb, *The Early Earth*, p. 98.

27. Whitcomb and Morris, *The Genesis Flood*, p. 69.

28. Morris, *The Beginning of the World*, p. 107.

29. Whitcomb and Morris, *The Genesis Flood*, p. 8, fig. 2.

30. Whitcomb, *The World That Perished*, p. 40.

8

Creationism: Strategy and Public Image

Creationist Challenges to Public Education

Christian creationists have been described as the religious right, fundamentalist, far right, conservative evangelical, ultra-conservative, literalist, and biblicist. Such groups generally agree that Bible-centered teachings must prevail in their schools. But what should be taught in the public schools?

For two centuries in America a constitutional barrier has discouraged overt religious indoctrination in public institutions. Such constraint has denied the certification of sacred writings as standard texts in public schools and the certification of teachers of religion who might interpret such writings.

Although public schools legally remain free of direct sectarian control, teachers driven by religious convictions may find covert ways to indoctrinate school children. Jerry Falwell has viewed the nation's schools as a large mission field and suggested a teaching strategy to guide students in specific religious rites:

> Over 36 million American school children are enrolled in public schools—a larger mission field than many countries. Can Christian teachers expose these students to the Gospel, or is religion taboo? . . . When the class studies Shakespeare's *Romeo and Juliet*, [the teacher] enlarges on the priest's description of God's grace and explains that grace is "God's riches at Christ's expense," and she gives step-by-step instructions on how to be saved.[1]

Obviously, legal restraints are no great hindrance to religious zealots. Despite constitutional law, creationists throughout America have long practiced religious indoctrination in public schools through prayer sessions, Bible reading and interpretation, and the singing of Christian hymns and songs.

A recent report indicated that about 42 percent of public schools in the South have regular sessions of spoken prayer.[2] The figure is about 15 percent nationwide. Smaller percentages of schools have periods for sacred songs and Bible reading. Such activities serve to indoctrinate and confirm students in sectarian beliefs, because teachers and students may freely address deities of a specific religious tradition. And, unfortunately, teachers also may be free to vilify opponents who promote allegedly false religion.

Fortunately, in some public schools fundamentalist religion versus other religion is not recognized as legitimate argument nor an essential part of the teaching program. And students are not pressured to embrace doctrines about scriptures, prayers, and deities.

Realizing that a strict religious approach cannot open doors to all schoolrooms, creationists have questioned what they might do to gain entrance and influence in the public schools. They have now found a clear answer.

Many creationist leaders have assumed a cloak of religion that includes a mask of science and are now asserting that they, in fact, are scientists. And the public education issue they push does not focus directly on religions but on two assumed models of science. Their model is "creation science." The other model is conventional science, which (except for parts that seem to bolster their position) they characterize generally as godless, evolutionary science.

Having assumed the trappings of science, creationists apparently are now determined to invite all citizens, particularly those in public education, to study and accept their two-model plan.

The Two-Model Formula: Creation versus Evolution

In 1974 Henry Morris and the staff of the ICR published the book *Scientific Creationism*. The centerpiece of that effort is their table of "predictions" concerning two models: creation and evolution. Their so-called predictions do not predict or foretell anything;

they suggest some condition or process relating to each key subject. The age of the earth, for example, is described correctly as being "extremely old" for the evolution model, but "probably young" for their model.[3]

Heading all suggested categories are natural laws.[4] Such laws in the creation model are said to be "invariable" but in the evolution model are "constantly changing." Both statements are false. I noted examples earlier, and in later chapters will give evidence that creationists deny constancy of natural laws. After naming galaxies, stars, other heavenly bodies, and rock types, creationists propose topics relating to the animate realm: life, organisms, kinds, mutations, natural selection, fossils, man, and civilization.

Two-model arguments in the above categories have been designed and used widely to challenge students, parents, educators, and other persons and groups concerned with public education. And just as energetically—but without great success—creationists have confronted legislative bodies and the courts. A prime target, however, has been and must continue to be the scientific community if they are to discredit currently accepted models of science.

Creationist Strategy: Maneuvers, Devices, and Gambits

Creationists aim to expose all of society to their brand of religion but for many reasons fail to reach that goal. A list of hindrances would be long but certainly would include a widespread indifference to the literal fundamentalist message. Added to the public apathy is a positive resistance from many groups, including mainline religious denominations, public service organizations, the news media, the bar and judiciary, and the scientific community. Creationists apparently have learned much in recent years from court decisions relating to public schools, specifically, that they cannot legally force themselves into positions of religious dominance.

Among the most obvious devices creationists have used to build prestige and discredit opponents is the selection of renowned antagonists (usually scientists) for display and debate in public forums. Such efforts at psychological conditioning require much energy and time and, apparently, are never staged primarily for audiences to learn science. The reports of such events by creationists often reflect favorably upon themselves (their personal style, their unsurpassed knowledge, and their triumphs) in contrast to the purported ineptness of their opponents.

Creationists attempt to orchestrate situations and manipulate people to achieve their goals. They are presently working diligently to change their public image from that of fundamentalist theologians to universal scientists. Thus, they presume to confront scientists on common ground. They also wish to establish that they and their opponents rely on belief systems—their opponents trusting mere theories whereas they possess biblical facts.

Creationists also would foster the notion that only two groups of antagonists exist (something like black and white pieces on a chessboard). Thus they manipulate all opponents into a single class. This false division greatly simplifies debating strategy and confers advantage of position. It also helps project the illusion of universal creationist power, because all adversaries of whatever kind must approach and contend with them. And having maneuvered every opponent to one side, they call them all evolutionists.

Because they cannot directly enter public schools as teachers of religion, creationists have made a bold maneuver, an undeniable gambit. As a chess player sometimes will sacrifice a pawn or other piece to secure advantage of position, so creationists seek advantage by suppressing sacred words. They do not discard beliefs and doctrines about God or scriptures or revelation; they only forgo the use of such terms in certain places. That commitment presumably helps to open doors that otherwise might remain closed. This gambit seems also to say that creationists have entered the halls of secular science and will abide by the logic and methods of science.

Eliminating sacred words, however, is only a preliminary move by creationists and part of a larger plan by which they hope to undermine confidence in their opponents and thus achieve their "scientific" (interpret: "religio-political") control.

Creationist Manipulations: Words and Catchwords

Two factors are necessary for communication. Humans first must agree on meanings of words, then must follow logical rules in bringing words together. For a rational transfer of ideas that leads to understanding, there must be agreement on semantics.

Unfortunately, slovenly and irrational use of words abounds, and educators are increasingly concerned about illiteracy and a general corruption of language. A related but more specific worry among scientists is the assault on basic scientific concepts. In their view, nothing is more threatening than the activities of those who would establish a system of religious authority invested with political power to dominate all scientific endeavor. The most insidious, and potentially most dangerous, are the endless strivings of religious fundamentalists who abuse the language of science to achieve that dominance.

Creationists, by their own testimony, see themselves in a universal battle for minds. Their goal is to spread religious ideas and indoctrinate society in fundamentalist values. To reach that goal, creationists have sought the most direct means to prepare minds and transform public opinion. What is their strategy?

Obviously, and I believe inevitably, creationists would first tamper with language—any language, ancient or modern—in ways to serve their purposes. This they have done repeatedly. For example, creationists misinterpret the biblical term "water" (Genesis 1:7) to mean water molecules. This is definitely an opportunistic approach to storytelling by avowed literalists who show little critical regard for the ancient text but, at the same time, declare the scriptures to be perspicuous. There is no biblical word that refers to vapor, the molecular state of water.

A second expression that creationists capture and put to work is the "greenhouse effect." To apply the term to alleged preflood times, however, they must turn it upside down and give it new meanings. Their massive greenhouse blanket allegedly brought only pleasant, universal warmth, a condition radically different from that forecast for only very small increases of heat-absorbing substances in the present atmosphere.

Another example is the fabricated hibernation story in which animals survived a year in Noah's ark. The Bible makes no reference to hibernation.

A fourth example is the opportune borrowing and abuse of the word "science." Added to creation it becomes the oxymoron, "creation science."

Is there explicit scientific meaning in the biblical word, "created" (Genesis 1:1)? The answer is a definite no. The biblical writers and translators have used the words "creation" and "create" as terms referring to supernatural acts of deity. There is no logical, understandable connection between the words creation and science. A leading creationist agrees, "We do not know how the Creator created, what processes He used, *for He used processes which are not now operating anywhere in the natural universe.* This is why we refer to creation as special creation. We cannot discover by scientific investigations anything about the creative processes used by the Creator" (emphasis in original).[5]

Although scientists agree that their investigations do not produce knowledge in a supernatural realm, they must reject Gish's logic in other statements. He begins with a negative certainty, "We do not know. . . ." Let us substitute his name and summarize as follows. Gish does not know about alleged divine creative processes. But at the same time he claims to know something about unknown creative processes. What he asserts is that the unknown processes, once operative, do not now operate anywhere in nature. He also affirms that no one can discover those unknown processes through any scientific endeavor. Gish expounds with vacuous certainty on a subject about which he admittedly knows nothing.

As for Gish's last sentence, how will scientists respond to affirmations of an invincible ignorance they cannot remedy? I believe they will agree. They know their research has never reached beyond the physical and experimental. Generally, therefore, they have turned from the mythical and unknowable to a cosmos of entities that sometimes yield to theoretical and empirical testing.

Much of science now enjoys a moderately high tide of public favor. Not surprisingly, creationists are determined to capture the word "science" and force it to serve their sectarian purposes. For some, merely using the word seems enough, as when apologists display a false devotion to disciplines of science but secretly work to promote religion. A remarkable short history of creationist debating and evangelizing strategies has been published by William Thwaites and Frank Awbrey.[6]

Fascination with Images and Masquerades

Creationist leaders are quite aware of theater and theatrics. To project widely a science-loving image, they seek the public stage, where they dramatically embrace "science." And with subtle moves and crafty use of borrowed words they appear to be what they are not. On occasion the actors put on effective masks or other disguises, for their success (before school boards, for example) may depend on concealing ideology and their true intentions.

An acclaimed consultant and expert in public relations and salesmanship has given written counsel to fellow creationists.[7] The primary focus of his article was on mistakes creationists make. Errors, he believed, could be traced to faulty methods in selling a "proper image" to school boards and other official bodies.

He emphasized repeatedly that school boards, legislatures, and courts "do not understand" creation (i.e., origins). Why this failure? Leitch explained: "The subject is too complicated. You cannot educate these boards with the tools and time available." And in his view, ". . . an educational effort is a waste of time."

If Leitch's statements are true, creationists have big problems. They must deal with a complex subject and with alleged creation illiterates plus their own confessed lack of tools and time. Creationists obviously need a simple, accelerated program. What might help to solve the problem?

Leitch strongly recommended that creationists sell what school boards understand: "Sell more SCIENCE."[8] But he cautioned, "A further suggestion, do not use the word 'creation.' Speak only of science. Explain that withholding scientific information contradicting evolution amounts to 'censorship' and smacks of getting into the province of religious dogma."

Leitch further instructed in the art of using labels, "Use the 'censorship' label as one who is against censoring science. YOU are for science; anyone else who wants to censor scientific data is an old fogey and too doctrinaire to consider" (capitalized words are Leitch's).[9]

Leitch also counseled fellow creationists to assist in locating up-to-date scientific textbooks. (From creationist-approved sources?) But he emphasized, "Through the whole process, assiduously avoid a creationist-promoting image." Obviously, external and visible appearance, the image, is crucially important for their purposes.

But under that masquerading image, what is the "science" that creationists are bound finally to advance? Leitch is positive: "We sell science that corresponds with Scripture."[10] That, however, is not the brand creationists were advised to display before school boards. The science for which they must make a spurious display is the science they secretly reject.

The masquerade raises disturbing questions. Do creationist leaders coach workers in covert acts to achieve hidden goals? Would they plan strategy in such a way that school authorities are not aware of what is taking place? Would they take advantage of school board members to establish curriculum goals? Would they in any way deceive by concealing their true motives? Is it possible that their program depends on deceit or concealment for success?

In summary, we have considered "pupils" being prepared with dos and don'ts for a stage-managed performance. They have been coached to "sell" what school board and other officials presumably understand and want. The selling is an ingratiating and subtle expedient. If their masks don't slip, the performers may be able to avoid exposure and achieve their goals.

Unfortunately, however, they may also be able to avoid the genuine interplay of inquiry and analysis that could provide a rational base for understanding and decision-making in public school policy.

Notes

1. J. Falwell (ed.), *Fundamentalist Journal* (September 1985).

2. T. O'Brian (New York: ABC News Report, March 23, 1987).

3. H. M. Morris, *Scientific Creationism,* General edition (El Cajon, Calif.: Master Books, 1974), p. 12.

4. Ibid., p. 13.

5. D. T. Gish, *Evolution? The Fossils Say No!* Public school edition (San Diego, Calif.: Christian-Life Publishers, 1978), p. 40.

6. W. M. Thwaites and F. Awbrey, "Our Last Debate: Our Very Last," *Creation/Evolution* 13, no 2 (1993): 1–4.

7. R. H. Leitch, "Mistakes Creationists Make," *Bible-Science Newsletter* 18, no. 3 (1980): 1–3.

8. Ibid., p. 2.

9. Ibid.

10. Ibid., p. 3.

9

Building "Creation Science"

Criteria and Structure

Gains in scientific knowledge depend on human perception and discovery in compliance with natural laws. In contrast, the fundamentalist creationists belittle human powers. The ultimate sources for their so-called science are biblical statements that, they insist, were revealed supernaturally.

Creationists do not call their viewpoint a theory. They would prefer to have people believe that they deal with facts, that is, fixed biblical knowledge, inerrant and unquestionable, revealed once and for all. Most fundamentalists therefore believe that mere studies and data collection never become crucial supports for biblical revelation.

Whitcomb stated that no single geologic phenomenon "demonstrates the universal flood."[1] Why is that so? He answered emphatically, "Scientific empiricism was never intended by God to be the direct and essential link to Biblical revelations."

Despite the ostensibly personal knowledge of what God "never intended," many creationists apparently believe that human observation and collection of data do have great value in support of Bible history. In fact, several creationist leaders see themselves capable of establishing universal flood and young-earth views from empirical evidence—without references to a God or other sacred entities.

Among the most ambitious and effective groups involved in

creationist lore is the Institute for Creation Research (ICR), headed by Henry M. Morris, president, and Duane T. Gish, vice-president. These leaders and others in the Institute have earned advanced degrees in the public educational system. In addition, they have been exposed to long training by rote and are thoroughly indoctrinated in fundamentalist Bible teachings.

Confirmation of the special preparation ICR members have received was publicized in an *Acts & Facts* article.[2] The author praised highly the fellowship and instruction offered in the Field Study Tour of the Grand Canyon and then reminded applicants about the instructors: "The ICR staff has trained and studied for many years to be able to apply the Bible model of earth history to the geology and biology of the area." That statement reveals a conformity to long training in biblical rote and the purpose of that training. The methods and motives are clear. Minds, thus prepared, will go out and gather data that plausibly can be made to fit biblical history.

In view of their commitments, I have summarized guidelines the researchers follow. The priority of activities will, of course, be changed by individual creationists, but the agenda comes from their own literature and promotions and clearly represents approved activities.

Research Guidelines for ICR

They will gather and report physical data that are in plausible agreement with biblical statements. The information sought will be at least superficially persuasive and must agree with the biblical interpretations and research goals of creationist leaders.

They will select research projects that have exciting elements of adventure and can be promoted as such. Creationists plan for spectacular, attention-demanding experiences, such as the ark search on Mt. Ararat, Paluxy tracks of dinosaurs and "people," an African dinosaur search (advertised but not yet conducted), and selected studies and adventures at Mount St. Helens and in the Grand Canyon.

They will selectively gather information to confirm creationist views—information that seems likely to frustrate and discredit opponents. The strategy is to prove opponents wrong and themselves correct. In contrast, I propose that virtually all scientists design experiments without ever considering whether their data will confirm or disconfirm biblical doctrines.

They will search continually for ways to discard long periods of time. Overshadowing the work of all young-earth creationists is the need to cut the ages of earth and celestial entities from billions down to a few thousand years.

Gathering support for Bible statements, scheduling spectacular search and adventure projects, hunting for data that may discredit opponents, and searching for any evidence that might compress time—such activities constitute much of creationism's field research. The following section reflects adherence to the above guidelines.

Creationism's Field Research

The ICR staff regularly conduct what they call empirical or field research. An *Acts & Facts* author outlined in detail the ICR commitment to "empirical" research by describing the projects given below.[3] Included are stated failures, complaints, results, or anticipated results.

Mt. St. Helens investigation. The purpose was to extend "documentation of the clastic dykes, microstratigraphy, rapid erosion, floating log mat, subsurface peat layer. . . ." Such data presumably would help creationists interpret "similar rock formations within a short-time catastrophic framework" (that is, in their preconceived framework of a one-year universal flood).

A short-time framework for rock deposition. Henry Morris stated that leading geologists (he doesn't name) have switched to the view that rock-forming sediment layers were deposited by different catastrophes over long periods of time.[4] In his view, this gave creationists a challenging project. ICR's commitment was definite: "ICR's job is to tie these layers together, showing that one continuous catastrophe was responsible for the majority of sedimentary rocks." "ICR's job," of course, is to force virtually all sedimentary activity into the single biblical flood year.

Outcrops at Split Mountain. Plasticity of rock layers allowed folding after deposition. This indicated to creationists that "no significant time had passed" after the first depositions. Eliminating time is a big part of their job.

Clastic dykes in Colorado Springs area. Studies of dyke genesis "seemingly eliminate about 400 million years of elapsed time." Also, in the Ute Pass, creationists assumed "no significant time lapse" in the uplift and folding of rock layers after deposition.

Outcrops in the Connecticut Valley. An ICR student completed a study of outcrops, and a supervisor "confirmed the many indicators of rapid processes present." Physical processes have to be rapid to fit creationism's flood scenario.

Yearly staff studies in the Grand Canyon. An ICR professor studied cement types in canyon wall strata. "Results are as yet inconclusive," said the author, "due to the limited sampling allowed."[5] In another project, an ICR student "has gathered data attempting to identify possible vertebrate footprints in the Cambrian strata. . . ." The motivation for this attempt is clear. One effect of such a finding would be to discredit the opposition who state that Cambrian deposits existed before the origin of vertebrates.

Three years later, an *Acts & Facts* report outlined research performed by ICR faculty.[6] Among their projects were studies of Carbon-14 and the atmosphere, the vapor canopy, and the ocean. The Paluxy footprint mystery was still taking up "research" time. One member studied Grand Canyon lava flows and also "returned to Mount St. Helens with a filming crew to study the rapidly formed deposits and eroded terrain from the air."

Creationists seem totally obsessed with problems of age and questions of time. They desperately search for evidence of rapid processes. Obviously, these workers anticipate that their findings will help confirm creationist views about early biblical history, specifically, recent creation, conditions in preflood times, and universal flood effects which took place rapidly.

Creationism's Aggravating Problems with Time

Although Francis Hitching has written several scientifically questionable books and received negative reviews,[7] he nevertheless should be given credit for exposing a very sensitive area of creationist thought. His probing interview with creationist Duane Gish raised important questions. The exchange was as follows:

When I asked him what were the biggest difficulties for creationist science—the points in a debate which he felt least
comfortable in answering—he answered after a moment's
thought that it was the apparently great age of Earth as shown
by the fairly recent advances in radiometric dating; and that
the fossil record could be interpreted as showing ecologically
complete ages—the age of invertebrates, the age of fishes, the
age of reptiles, and so on up to the present.[8]

Hitching then asked whether the "great age" problem was ever
serious enough to cause doubts about biblical truth. Gish answered
with this apology: "No, not at all. It just means we've got to get
to work on the science of these things and find out why they
do not confirm the Biblical account more readily." Gish's answer
seems to reflect the general reaction and attitude of many
creationists: keep working hard; try to learn why the earth and
universe appear so much older than Bible writers represent them.

Opponents can take Gish's reply as an implicit confession of
mental frustration. Moreover, opponents can appreciate that
creationists have never seriously gotten into "the science of these
things" (fossils, ages, epochs).

Why do creationists find that scientific data about time fail
to agree with their religious beliefs? A reasonable answer may
be taken from their experiences noted above. As inveterate
opportunists constrained by scriptural bias and rigid agendas,
they scurry about looking for proofs of what they already have
decided are facts. How defensible is the position of those who
always look to confirm what they believe, but conduct no tests
that might falsify a favorite belief? One may reasonably wonder
how devastating is the bias, or deep the psychosis, that will not
allow partisans to probe their honest suspicions, nor let them
ever believe that at some stage of training they could have been
misinformed.

Creationism's Library Research

Although the Institute for Creation Research conducts numerous
"field" projects, the Institute places primary emphasis on publications. To advance that program, ICR personnel depend largely

on what they call "theoretical or library" research. Henry Morris explained several important criteria and practices employed in the ICR literature search and publication program.[9] First, creationists recognize that massive amounts of experimental data are available for analysis and reinterpretation from their viewpoint. They also find virtually unlimited quantities of reliable data—observations—that need no further validation but need "only a new theoretical framework."

"Accordingly," Morris disclosed, "all the ICR scientists are continually reviewing data published in their own fields which might be relevant to scientific Biblical creationism or catastrophism, and then reinterpreting them in that context."

Morris also mentioned another data-search category that has very high priority: newly developed studies that disclose "processes which can be applied as geochronometers to support the Biblical doctrine of recent creation."[10] Obviously, creationist religious beliefs and goals dictate search and review patterns, specifically for data "which might be relevant" or "which can be applied" in support of biblical doctrines. High on their agenda are any data that seem to eliminate large blocks of time.

Among the most visionary projects to which creationists are now committed is the complete revision of the standard geologic column. Their goal is to make it conform to their creation and flood model. Standard geologic charts generally represent historical time as eons, eras, periods, or epochs and, within those time frames, list important changes in earth and living things. Such records of events are the legacy of many thousands of workers in geological, biological, and related sciences. Such masses of data, creationists have concluded, are just waiting for their cut-and-paste operations. The procedure is to select a block of data—for example, the stratigraphic features and chronology of an era—then transfer the block into some narrow temporal framework of biblical history.

Creationism's block-and-move operations can be illustrated as follows. The Azoic (the "without life" or Precambrian) Era, in which earth was formed billions of years ago, must be compressed into an alleged creation period of two days plus some part of a third day (Genesis 1:1-11). By strict fundamentalist reckoning, the creation week took place within the past 10,000 years. Evolutionary scientists consider the Azoic Era to be a long accretionary period in which earth grew to about its present size. Creationists also move the Archeozoic Era into the biblical week of creation. And most of the remaining earth history (Proterozoic to Tertiary) they squeeze into the legendary one-year period of the Genesis flood.[11]

Detailed manipulations of geological time have been published by John Morris in his chart of the geological column.[12] He moved the entire Paleozoic Era, nearly 400 million years (which he mistakenly identified as Cenozoic), to somewhere within the first forty days of the Genesis flood. On his chart, the forty days would extend into the Triassic and perhaps the Jurassic Period. John Morris's flood year extends from the Cambrian into the Tertiary Period of the Cenozoic Era (mistakenly identified as Paleozoic). He thus compresses a vast sweep of time—nearly 600 million years—into a single flood year, which allegedly occurred about 4,300 years ago. The ratio 600,000,000 to 1 is indeed a remarkable compression of time.

An even greater compression appears in Morris's division of the Paleozoic Era. He cut out a block of Cambrian, Ordovician, and some Silurian time (the sum equals about 175 million years or 63 billion days) and made it equivalent to alleged "Early phases of the flood. . . ." Any number of days less than thirty-two, taken to represent an early flood phase, would give a time compression ratio in the billions.

Such ratios, however, should not be surprising. Creationists have lowered ancient mountains. They have elevated high ranges. They have raised ocean floors and have collapsed them in the same year. Manipulations of time and matter are indeed unchanging methods for inveterate builders of the upside-down pyramid.

Notes

1. J. C. Whitcomb, *The World That Perished* (Grand Rapids, Mich.: Baker Book House, 1973), p. 144.

2. H. M. Morris, "Grand Canyon Study Tour," *Acts & Facts* 18, no. 2 (February 1987): 3.

3. H. M. Morris, "ICR Field Research, 1986," *Acts & Facts* 15, no. 11 (November 1986): 1, 5.

4. Ibid., p. 1.

5. Ibid., p. 5.

6. H. M. Morris, "ICR Field Research," *Acts & Facts* 18, no. 11 (1989): 6.

7. R. Dawkins, *The Blind Watchmaker* (New York: W. W. Norton, 1987), pp. 79–81, 84; T. McIver, "Verna Wright; Francis Hitching," *Creation/Evolution Newsletter* 7, no. 5 (1987): 15, 16.

8. F. Hitching, *The Neck of the Giraffe or Where Darwin Went Wrong* (New York: Ticknor & Fields, 1982), pp. 15–16.

9. H. M. Morris, *A History of Modern Creationism* (San Diego, Calif.: Master Book Publishers, 1984), pp. 254–55.

10. Ibid., p. 255.

11. H. M. Morris, *Scientific Creationism*, General edition (El Cajon, Calif.: Master Books, 1974), p. 129.

12. J. D. Morris, *Tracking Those Incredible Dinosaurs and the People Who Knew Them* (San Diego, Calif.: CLP Publishers, 1980), pp. 56–57.

10

Those Credible Dinosaurs
That People Never Knew

The Paluxy Mantrack Wonders

To those yearning for signs and wonders, the short article by
Clifford Burdick was a marvelous revelation.[1] How could the
credulous doubt Burdick's story? He told of giants whose footprints
were still visible alongside dinosaur tracks in a central Texas
riverbed.

The mantrack fantasy had its widest exposure (now, more
than thirty years) when Burdick's photographs were published
in *The Genesis Flood*.[2] Legends to the photographs define them
explicitly: "Figure 10. CONTEMPORANEOUS FOOTPRINTS OF MAN
AND DINOSAUR" and "Figure 11. GIANT HUMAN FOOTPRINTS IN
CRETACEOUS STRATA." Creationists borrow geological terms freely
but apply new meanings. Their "cretaceous" is within the past
10,000 years.

Among the most effective means of spreading the mantrack
myth has been a Films for Christ movie produced by Stanley
Taylor. Henry Morris described Taylor's embrace of "strict cre-
ationism through reading *The Genesis Flood*" and explained how
Taylor was "especially intrigued" by Burdick's photographs in
the book, showing both human and dinosaur footprints in the
Paluxy riverbed.[3] Stanley Taylor released a film in 1971 called
Footprints In Stone, which, by Morris's estimate, "has been seen

by over 600,000 students in public school showings and by over one and a half million people around the world." Morris stated that Films for Christ has withdrawn the Taylor movie from circulation.[4]

As it has turned out, the imagined footprints that Stanley Taylor and others promoted never occurred in stone; they were products of undisciplined imagination. Unfortunately, the films and photographs and extensive literature on the mantrack myth have contributed their part to the present illiteracy in science.

The most ambitious of the creationists who have written about Paluxy mantracks is John Morris,[5] who also published Burdick's photographs. Much of his book focused on information he thought pertinent to his central assertion, "that man and dinosaur walked at the *same time* and in the *same place*" (emphasis in original).[6]

John Morris had virtually unlimited hopes for the success of his book. He quoted Wilder-Smith, who had confirmed another scientist's opinion, "that a single such find [coexistence of man and dinosaur] would provide sound reason for renouncing all evolutionary theory."[7] Responding to that pronouncement, John Morris informed his readers, "The purpose of this book is to provide such evidence."[8]

His purpose and strategy apparently matched his tantalizing vision. John Morris would document evidence and gather "all worthwhile information" into one source book. His goals were to encourage those who accept his data and conclusions, confront those who don't, but provide for everyone his "comprehensive reference tool."[9]

John Morris has had some success in realizing his first two goals. But unfortunately for creationists, he and his colleagues have not moved one step toward the third goal. Proof of this comes from written testimony by creationists.[10] We learn that John Morris's book was not comprehensive and certainly not definitive. His disclaimer confirmed that his book could not be relied on to present a true picture of Paluxy footprints.[11]

John Morris's book continued to be advertised and sold into the late 1980s, although it certainly was not a work of science. The central theme did not conform to the reality of Paluxy. And most embarrassing, its author and the ICR staff had to deny its integrity. In the first place, the title was not authentic; it was an apt selection for fabrication. Moreover, dinosaurs are not, and never were, incredible; they left credibility in every footprint. Most devastating was the fact that John Morris never tracked "people who knew" dinosaurs. The "people" were illusory and incredible— not the dinosaurs.

Creationism's Paluxy Fiasco: Model of Pseudoscience

Respected scientists embrace objectivity and bow to scientific methods. I review how creationists were led astray and into the embarrassing Paluxy episode in Texas. Their own literature provides the evidence.

They listened to anecdotal testimony. Long-time residents allegedly had seen "true" human footprints in the Paluxy riverbed, but, alas, the prints had become badly eroded, or completely eliminated by floods, or had been removed in slabs of limestone and sold. Listening to "old ones" who delight in their anecdotage— but not in science—can be a risky temptation.

They believed old-timer's explanations about holes in the Paluxy streambed. Some creationists seem to have been fascinated and overcome by these excavations. John Morris concluded, "The [Taylor] trail might not have been called human if not for the hole from which a 'perfect' print had reportedly been taken some 50 feet away."[12]

They disregarded problematic features of tracks. Unquestionable human toeprints were absent. Other diagnostic features were absent. And the decisions about human prints and tracks that should have been uncertain and tentative (because based only on general notions of outline, shape, and stride) were presented as clear and settled.

They rejected the cautious approach and conclusions of other investigators. John Morris noted that problems with tracks prompted creationist Berney Neufeld to label the Taylor Trail impressions as "shallow, eroded dinosaur tracks."[13] Glen Kuban, also a creationist, explained that before Taylor's film was released in 1971, a team of creationists from Loma Linda University had studied the Taylor site and expressed doubt about human footprint claims.[14]

They failed to scrutinize pertinent data in their possession and in the field. For example, an *Acts & Facts* report disclosed that, "over 100 clear dinosaur tracks have now appeared . . . [as stains showing tridactyl reptile patterns] which had never before been discovered."[15] And in the same paragraph was this rather oblique admission, "some hints of these dinosaur toe stains have possibly been discerned on photos taken when the prints in question were originally discovered." These statements indicate that pertinent riverbed and photographic data were available early, but were not analyzed carefully enough to avoid prejudice and serious errors in publications.

They succumbed to unyielding personal biases. John Morris and other creationists brought mental commitments to their early data gathering. This largely accounts for serious errors in their interpretations and writing. In their Paluxy experience, apparently everything they saw, heard, touched, or thought about had to be forced into a single mold. And in that mold, humans and dinosaurs were always contemporary.

Kuban's update and assessment concerning the assumed co-existence of dinosaurs and humans was pessimistic.[16] Creationists were still searching for evidence. Concerning the Paluxy controversy he stated that, "The once-clearing waters . . . are now unnecessarily muddled again. . . . Evidently little if anything was learned from past mistakes."

Creationism's Juvenilities: Dinosaur Pets and Servants

The Paluxy mantrack fiasco took its toll on the momentum and credibility of the creationist movement. The footprints that Stanley Taylor and others promoted were never found in Paluxy riverbed stone; they were fabricated in their own minds.

Creationist storytellers and publishers, however, do not easily give up their fantasies about humans and dinosaurs living together and walking the same trails less than 10,000 years ago. Further, they are determined to change radically the public perception about the nature and behavior of dinosaurs. Their propaganda and illustrations depict dinosaurs and people in common and cozy relations—something like the animals now being kept for children's enjoyment in petting parks and zoos.

During the early-to-mid-1980s, when scientific findings[17] were discrediting the creationist Paluxy story, Paul Taylor, son of

Stanley Taylor, was preparing materials for a new "Christian" film and a new book, *The Great Dinosaur Mystery and the Bible.*[18] Anonymous sales promotions in the July 1987 *Acts & Facts* predicted that the book was "On its way to being the best selling Christian children's book in 1987!"[19] The advertisement made this declaration and an implicit promise of truth: "IT'S TIME TO TELL THE TRUTH ABOUT DINOSAURS."

What is the "truth" about dinosaurs that creationists wanted children and adults to learn? The glossy, colorful pictures on the outside front and back covers of the book, viewed together, tell the basic story. A child stands near a pond watching animals a few feet away: a kudu, a bird, three dinosaurs, and an elephant. The author selected all vegetarians—not a meat eater in sight. Pages 20–21 depict a parrot and three dinosaurs, also herbivores. And a man embraces a dinosaur, their faces only a few inches apart. A two-page spread (pp. 54–55) again pictures herbivores: a rabbit, beaver, deer, two dinosaurs; and a man strokes the head of a dinosaur.

A special note by Henry Morris on the inside cover described the book as "wonderfully unique" and "fascinating" for children. He also stated that the book contributed to biblical apologetics, so that even adults would profit greatly from reading it.

In addition to contrived artwork, what "truth" did Paul Taylor choose to publish? The format he adopted presents a system of religious training that can be used for rote memorization and recitation—a perfect catechism. I present several of his questions and answers, and some questions of my own.

1. "Why did God create dinosaurs?" All animals, including dinosaurs, were for the benefit of humanity.[20] (All animals? Fleas, ticks, leeches, internal parasites?)

2. What benefits did dinosaurs provide? In general, they helped control lush plant growth in forests. Taylor explained that long-necked dinosaurs could have eaten the upper foliage of tall trees to bring light to understory plants.[21] Dinosaurs could also have "cleared paths through the forest." (Everyone should know that man needs to get around in the forest.)

3. "Why did God invent so many different kinds of . . . animals? Perhaps because He wanted to delight Man with His power, wisdom and love."[22]

4. What reasons did God have for creating very large dino-

saurs? One purpose was "surely to impress Man." They would show the Creator's great power.[23] Other reasons noted above were practical services: tall tree pruning and creation of forest trails.

5. What was dinosaur life like in the beginning? Adam and Eve ruled over all animals in a perfect environment where every type of plant food existed. All animals were friendly. They didn't eat meat or kill. There was no death or disease.[24]

6. "Where did most of the dinosaur fossils come from?" Simple answer: "Noah's Flood."[25]

7. "Did Noah take dinosaurs on board the ark?" God brought them to Noah, probably, "a young pair of each main type of dinosaur . . . basic kinds . . . not every variety that had developed since Creation."[26]

8. "After the Flood, what happened . . . ?"[27] Dinosaurs lived a few centuries, never in large numbers. "Mud and rock were everywhere. Cold, forbidding mountains stood where none had been before." Temperatures were extreme, much hotter; deserts appeared. Snow fell for the first time in some places. "A short ice 'age' followed the Flood. . . ." Taylor had the catechism all worked out for children and impressionable adults. Obviously, he didn't have to bother with evidence for the above statements in attempts to mold thought patterns in a child. Unfortunately, some children exposed to such training may never recover.

Taylor takes up considerable space discussing "dragon" legends, which, he says, "must have come from memories of dinosaurs."[28] The point that Taylor makes throughout his book is that humans and dinosaurs were contemporaries. He then comes to the question of meat-eating.

9. "Did animals eat meat *before* the Flood?" Taylor emphasized that killing for food "could not have started until sometime after Man's first sin."[29] He thought it unlikely that such habits were acquired instantly. "And," he conjectured (concerning meat eaters), "they may not have been common until sometime after the Flood." Moreover, he surmised, "By the time of the Flood, most animals must still have been able to live on plant foods alone."[30] He then stated dogmatically, "Every basic kind of land animal and bird in the world was on the Ark. They ate only plants during the voyage."[31]

The exclusive plant diet is definitely another basic "truth" that Taylor must get across to children and adults. Taylor emphasized that on the ark the stored foods that "the animals would need . . . must have been . . . 'every green plant.' "[32] Without miracles, green plants would soon become dry hay. One can only wonder how such a diet would suit herbivore hummingbirds, snakes, and the toothless anteaters.

10. Were dinosaurs ever as ferocious as they are depicted? Taylor surmised that dinosaurs "must have been harmless— designed to delight man and benefit the world, just like all the other animals."[33] He noted that, "New research suggests" the following: Tyrannosaurs were unable to move quickly, and most other dinosaurs could easily have avoided them; their teeth were not well rooted; and their front legs seemed too weak for grasping and killing.[34] How valid are these four objections? Would tyrannosaurs have to move quickly to secure adequate food? Did they need to run down the swiftest animals? Would alleged tooth problems or weak front legs mean that they were not effective predators? Taylor's "new research" suggestions are not convincing.

11. How were dinosaurs' horns, claws, and other specialized structures used? Claws, head horns, rows of heavy spines, clubbed and spiked tails, heavy armored plates, and bony frills at the back of the skull are not mentioned by Taylor as being useful for defense. Why? The reason, apparently, is that such information would introduce predator-prey concepts, which would contradict his animal stories and herbivore notions.

12. How did tyrannosaurs use their claws and teeth? Taylor suggested, "Perhaps . . . to tear up tough plants and fruits, not dinosaurs."[35]

13. How did dinosaurs like *Triceratops* use their head horns and teeth? Taylor guessed that "The head horns may have been used for getting food by lifting thick foliage. They could be used for poking, rooting or turning plants. Dinosaurs like the *Triceratops* had very strong jaws and replaceable teeth. Using their extremely strong jaw muscles, they could slice through very tough plants— even good-sized branches and roots. Dinosaurs like this could even have chewed on tree trunks."[36]

Obviously, Paul Taylor tried very hard to depict all dinosaurs as docile, completely tractable servants of man—and, of course, vegetarians.

Creationists do not give up easily their ideas about humans and dinosaurs living together 4,500 years ago. Their strategy is clear: What they lose at Paluxy (imaginary coexistence of man and dinosaur) they bring back in more sweeping generalizations, but they offer no specific sites or other verifiable evidence that anyone might investigate directly. Further, they withhold their writings from impartial review of scientists and convey them to children and adults who lack the critical ability to recognize religious fabrications.

The Paluxy mantrack fantasy and other stories about dinosaurs and people clearly illustrate how zealots blend artifact and illusion in speculative field research, film making, and writing. It is most disheartening to see how easily untruth can spread into the homes and schools of America and around the world.

Notes

1. C. L. Burdick, "When Giants Roamed the Earth," *Signs of the Times,* July 25, 1950, pp. 6, 9.

2. J. C. Whitcomb and H. M. Morris, *The Genesis Flood* (Phillipsburg, N.J.: Presbyterian and Reformed Publishing Co., 1961), pp. 174–75.

3. H. M. Morris, *A History of Modern Creationism* (San Diego, Calif.: Master Book Publishers, 1984), pp. 283, 285.

4. H. M. Morris, "Following Up on the Paluxy Mystery," *Acts & Facts* 15, no. 7 (1986): 7.

5. J. D. Morris, *Tracking Those Incredible Dinosaurs and the People Who Knew Them* (San Diego, Calif.: CLP Publishers, 1980).

6. Ibid., p. 4.

7. Ibid.

8. Ibid.

9. Ibid.

10. J. D. Morris, "The Paluxy River Mystery," *Impact,* no. 151 (1986), pp. i–iv; H. M. Morris, "Following Up on the Paluxy Mystery," p. 7.

11. Morris, "The Paluxy River Mystery," pp. i–iv.

12. Ibid., p. ii.

13. Ibid.

14. G. J. Kuban, "A Summary of the Taylor Site Evidence," *Creation/ Evolution* 6, no. 1 (1986): 11.

15. Morris, "Following Up on the Paluxy Mystery," p. 7.

16. G. J. Kuban, "Retracking Those Incredible Man Tracks," *Reports* 9, no. 4, 1986 (Special Section, unnumbered four pages) (National Center for Science Education).

17. Six articles and a bibliography of 114 references make up a special issue that solves the Paluxy "mystery." See J. R. Cole and L. R. Godfrey (eds.), "The Paluxy Mystery—Solved," *Creation/Evolution* 5, no. 1 (1985): 1–56.

18. P. S. Taylor, *The Great Dinosaur Mystery and the Bible* (El Cajon, Calif.: Master Books, 1987).

19. Anonymous advertisement, July 1987, p. 8.

20. Taylor, *The Great Dinosaur Mystery and the Bible,* p. 20.

21. Ibid.

22. Ibid.

23. Ibid.

24. Ibid.

25. Ibid.

26. Ibid., p. 32.

27. Ibid., p. 34.

28. Ibid., p. 36.

29. Ibid., p. 51.

30. Ibid.

31. Ibid.

32. Ibid.

33. Ibid., p. 56.

34. Ibid., p. 57.

35. Ibid., p. 58.

36. Ibid., p. 59.

11

Genesis Flood Rewrite

Religious Exercises and Intellectual Difficulties

Leading creation fundamentalists perform several remarkable services (ostensibly necessary) in society. One is to impress on everyone that rejection of the flood record in Genesis stems from intellectual and moral pride of rebellious sinners and reflects a religious problem, not a scientific one. Thoughtful persons among the condemned, however, may not be intimidated by such talk. The insult is not in being called sinners; it is the notion that rational beings cannot possibly have intellectual problems when they read the Genesis flood narrative.

A second remarkable practice among religious fundamentalists is the lip service they pay to the omniscience and power of deity. Besides making partisans feel comfortable, this ploy seems to reflect proper faith and confirm humble servants in orthodoxy. With such preliminaries out of the way, servants may then go about developing ideas that deny the declared attributes of deity.

In the introduction to their book *The Genesis Flood*, Whitcomb and Morris professed full belief in the divine inspiration and verbal infallibility of scripture.[1] They affirmed, moreover, that "the Creator is also the true Author of Scripture." These authors apparently stand on the solid ground of faith and orthodoxy.

Nevertheless, some creationist leaders have had difficulties with the flood narrative. Was it a religious problem, one that raised uncertainty about the power and authority of God? Or was it a scientific problem, centered on questions of intellectual integrity? Three Bible verses (Genesis 6:13, 21, 22, [RSV]) pertain to these questions and raise basic issues. The narrative is detailed and explicit: "And God said to Noah . . . . 'Also take with you every sort of food that is eaten, and store it up; and it shall serve as food for you and for them.' Noah did this; he did all that God commanded him."

In those simple expressions is a literal message the biblical author meant to convey: The divine providence prescribed and commanded the collection and storage of food because of its vital role in preserving life.

Changing the Biblical Plan

Creationists Morris and Whitcomb apparently chose to ignore the subject of food. They have not emphasized that stored provisions were necessary to preserve life. Instead, they developed another story. It was a survival tale based on a different plan.

Henry Morris speculated on the "remarkable physiological mechanism" that works to "suspend all bodily functions in the state of hibernation."[2] He expounded on "divinely-ordered genetic mutations" imparted to selected animals, which prepared them for migration to the ark and a year's hibernation in the ark. He concluded with impressive rhetoric,

> Then as they arrived at the Ark, and entered, and in response to the suddenly darkened sky and the chill in the air, they settled down for a year-long "sleep" in their respective "nests" in the Ark.[3]

John Whitcomb also stated that an imposed hibernation reduced life processes to a minimum.[4] He then added this remarkable conclusion: ". . . and thus removed the burden of their care completely from the hands of Noah and his family." That bold statement implies that in their alleged flood-year many thousands of animals were completely without management and feeding care. Whitcomb further defended his hibernation surmise,

. . . even if only domesticated animals were taken on board . . . it *still* would have been a gigantic if not an impossible task for eight people to care for *hundreds* of animals in a floating barge for months![5] (Emphasis in original)

Morris and Whitcomb have revealed unique, fabricated aspects of the Bible story in the statements above. We understand the intellectual problem and their solution. The labor shortage was severe, and the work load impossibly large. In the minds of Morris and Whitcomb, hibernation was the only solution. No less remarkable than the storytelling was their unscientific methods. A bit of surmise clothed in specious rhetoric automatically became "creation science."

Biologists who deal with living things and vital processes may well reject both plans—feeding care and hibernation—recognizing them as formulas for death. They understand Whitcomb's concern about too much work and too little help. But they will reject irrational fabrications about a year-long hibernation of animals and their subsequent dispersal over the whole earth.

The hibernation-survival plan cannot be accepted without proof of the following assumptions. All animal species hibernated for a whole year. They survived without feeding. Despite the destruction of food reserves by a universal flood, starved animals released from the ark spread abroad as viable breeding pairs. Moreover, they all found nutrient and energy sources needed for population increase, despite too little time for resource recovery. I doubt that competent physiologists and experts in animal behavior and care will believe such experiences could occur.

Reactions to the Rewritten Story

Public reactions to the rewritten Bible story probably range from credulity to critical unbelief. Uncritical readers might marvel at the alleged ark experiences. Perceptive literalists will be skeptical. They know that the Genesis account says nothing about hibernation. Further, they must be suspicious of anyone who rejects the explicit word of God. The strict biblicists could declare that Morris and Whitcomb subvert scripture and question God's ability to make wise decisions. (These authors nowhere state that God's food plan was defective or poorly conceived; it was merely unworkable.)

Nor will believers inclined to be skeptical embrace the revised story. For them the narrative is at best an allegorical tale, a tribal myth. They will see the tedious explaining as a waste of time, as if rhetoric could transform the myth to fact.

Nonbelievers among scientists and science writers will reject the flood story as fiction and conclude that creationists who both defend and revise it are confused and unreliable.

The most critical opponents will view creationist authors as pretentious and condescending. Professing belief that the Bible was divinely inspired (even to choices of words by human authors) and therefore infallible, these same professing believers have rejected the biblical food-for-survival plan and substituted their own. Critics should realize, however, that creationist planners did allow deity to go along with them. In fact, they were disposed to let deity be the author of their plan, stating that the hibernation was supernaturally imposed.[6]

Clearly, leading creationists have not been concerned with the nutrient needs of animals in the alleged ark experience.

The Flood Myth Improved?

A final question pertinent to the flood problem is this: How can one improve myths? Admittedly, some myths describe imaginary beings who are honorable and wise. Some are senseless; others are silly. Still others convey images of extreme cruelty. To "improve" a myth, one might delete or revise parts that are offensive. For example, a fierce god who devours his own offspring might be limited to other flesh, possibly his enemy's children, or perhaps be transformed into a staunch vegetarian. But despite the revision, the myth does not become historical fact.

Animal care on the biblical ark would have been an impossible task. This raises disturbing questions and doubts about the story, and in the minds of some would indicate a mythical origin. Rewriters Morris and Whitcomb apparently decided that an imposed hibernation would obviate the need for a massive labor force on the ark. (No one would need to call in angels to do the work.) And the hibernation, if worthy of belief, would also enhance the story's credibility.

Over the centuries, the Genesis flood has been a great challenge for writers and preachers, especially those who would confirm alleged flood events as historical truth. In this regard, modern creationists have been preaching and writing extensively over the last quarter century about imagined implications of the biblical flood for biology, geology, and paleontology. Not unexpectedly, an explanatory myth, such as the flood narrative, becomes incredibly complex in the hands of imaginative interpreters and writers. All the feverish expounding and explaining may fill libraries, but the myth never turns into fact. It is forever myth.

Notes

1. J. C. Whitcomb and H. M. Morris, *The Genesis Flood* (Phillipsburg, N.J.: Presbyterian and Reformed Publishing Co., 1961), pp. xx, xxii.
2. H. M. Morris, *The Beginning of the World* (Denver, Colo.: Accent Books, 1977), p. 98.
3. Ibid.
4. J. C. Whitcomb, *The World That Perished* (Grand Rapids, Mich.: Baker Book House, 1973), p. 32.
5. Ibid.
6. Whitcomb and Morris, *The Genesis Flood*, p. 74; Whitcomb, *The World That Perished*, p. 32; Morris, *The Beginning of the World*, p. 98.

Part Two

Contrasts and Challenges
Science versus Creationism

12

Building Science
Criteria and Structure

Natural Foundations and Human Discovery

Scientific knowledge rests on a base of natural laws, and gains in such knowledge arise from rational processes of human discovery. But that is not all. Science depends on processes that conserve and communicate knowledge arising from discovery.

The great discoverers have set in place the physical groundwork and cornerstones for a massive repository of science, a unique structure that increases in size and complexity but is never finished.

Those who contribute to that structure have received training and help from many sources, but they must quarry their own stones of verifiable facts. The most valued facts, personally quarried and shaped perhaps to be laid systematically along some rising wall, are not set in place without a mortar of rational hypothesis or theory.

If building stones are verifiable facts, then theory is the mind's attempt to explain and tie facts together. Theory is the bedding mortar that provides rational coherence and binds the stones in one course to those in lower courses as well as to new stones eventually to rest in upper tiers.

The ideal theory sets up and hardens around all relevant facts and becomes part of a useful, unitary structure. In some instances, the mortar of theory seems to endure weathering and other critical stresses as well as the rock-hard facts, so that together they display properties of a monolith. Thoughtful and objective scientists, however, will routinely inspect the mortar—whether old or new— to detect flaking or other disintegration. Deficient theories are expected to crumble and may need reformulation or replacement, but valued facts remain.

Scale of Comparisons: Guessing, Surmise, and Theory

Readers may have noticed in earlier chapters the use of such words as guess, surmise, or conjecture—but never theory—to describe creationist ideas. My intention was to present accurately the basic ideas and related assumptions held by creationists. Scientists do not approve of guessing as a reliable means of acquiring knowledge. The reason is that a sheer guess depends on chance for verification and not on a rational process of observation, hypothesis, and experimental testing.

Propositions born of surmise or conjecture are not much better than guesses. Proponents of such give little or no factual information, and their "evidence" may be so trifling that it is given with a clear awareness of deficiency.

This defines the dilemma of some present-day creationists who admit that they have no verification of their vapor canopy's existence. Society, therefore, may reasonably judge creationism's vapor story to be nothing more than surmise. To speak of canopy theory is too generous. A search of fundamentalist literature reveals nothing that would raise the canopy notion to the level of theory.

Advanced societies provide opportunities for human discovery and progress in science. Language and information systems in such environments allow youth to develop early their literacy and learning skills. Thus, individuals observe and organize and, if given freedom, begin to assess ideas critically. Fortunately, children reach a stage at which they want to know, "Is the story true, or is it only pretend?" With greater maturity, some will extend their mental probing to the assumed facts of dominant folk science and other lore, classifying some perhaps as products only of superstition and myth.

Offering surmise and conjecture to those who yearn for secure knowledge can only frustrate and disappoint. Enquiring minds reach for standards at higher levels, as the reasoned hypothesis (at least initially), and may finally come to rest more or less securely on accepted theory. Two concerns must remain important for scientists. The relative position of theory among other concepts must remain high, and theory must agree with natural laws. The term theory is not a synonym for "guess," as illustrated in the disparaging remark: "It's only a theory."

Primacy of Earth's Life-Support System

A coherent model of the history of living things and their fossils must require favorable conditions for life's beginning and continuation. This is a self-evident truth that leading fundamentalist creationists seem to forget. Or it may be that they don't understand—or they are unwilling to consider seriously—the deadly effects of strange meteorology, geology, and oceanography they have claimed for the primitive world.

It is clear that, whatever their state of mind, creationist leaders routinely advance arguments that avoid or obscure basic scientific issues. They employ abstract statistics zealously, but never enlighten people on the probability of life existing under an invented vapor canopy that would have blocked light from the earth. Have any creationist physiologists or chemists dealt with the problem of light blockage that eliminates life-supporting photochemistry of plant pigments?

Creationists argue about gaps in biological fossil records— an argument completely irrelevant, if their world could not have supported life. Futile also are the arguments about metamorphosis of butterflies on a plantless earth.

Creationists parrot laws of thermodynamics, especially the Second Law (the entropy law). They then invent scenarios that deny operation of that law in their preflood world. For example, they abolish winds and, thereby, ocean currents. And then, incredibly, they preach about a universally warm and seasonless planet. Where are the dynamics of energy flow when heated air and warm oceans could not transfer energy poleward? Consider also the mistaken thermodynamics where insolation continues and heat builds up for centuries under a vapor canopy without a balancing heat loss to space.

Creationists also deny energy dynamics (the Gibbs free energy) necessary for animal survival. Are there any animal physiologists who will defend the fabrication that pairs of every land animal could have hibernated on a mythical ark for 371 days without receiving food energy? Indeed, creationists have disregarded natural laws and physical needs and have created scenarios of death.

Strategy of Scientific Rebuttal

Not only do creationists challenge science over a broad front ranging from astronomy through the earth sciences, they confront opponents on subjects in the life sciences and in the allied field, paleontology. The decisive issue in these challenges is that beliefs held by creation "scientists" are based on their interpretations of sacred writings, not science. Their primary goal is to advance fundamentalist religion.

It is time for scientists to insist that creationists defend their glib assumptions about physical conditions on the primitive earth. Those assumptions have to do with water, air, light, and heat. The fifth essential among the primary constituents that support life are mineral elements (discussed in earlier chapters). Obviously, these five entities existed before life.

It is obvious also that facts and events which pertain to life (heredity, variation, morphology, speciation, fossils, etc.) are dependently related to primary entities noted above. Although I have avoided long discussions of genetics and related studies, no one should minimize their importance. Indeed, such topics deal with physical reality of living or once-living things and represent fields of valuable research.

The more basic considerations, however, are those elementary factors that existed before living matter and constitute earth's life-support system. It is here that creationists have made huge mistakes.

Outline of Determinative Issues

Some creationist beliefs were first illustrated as stones in the inverted pyramid (see fig. 1, p. 19) and later stated briefly in chapter 1. Here I will outline objections to show that "scientific creationism" is not scientific. Brief discussions give clues to more detailed examinations in later chapters.

Dominance of Religious Faith. By sheer acts of faith, biblical literalists believe that creationism's water shell was lifted above earth at creation. Because the water shell's existence has not been verified (as creationists admit), the mere belief that it did exist cannot be dignified as science. The issue here is the dominance of religious belief. It is determinative because it prescribes the domain in which creationists operate. For them, every aspect of science ultimately must be subordinate to religious faith—an obsession that inevitably hinders the intellectual progress of society.

Water Assumed to be Vapor. The creationist assertion that a vaporous water shell enveloped earth has been accepted as truth by literalist believers, despite serious contradictions. The Bible has words for liquid and solid states of water, but no word for vapor molecules. Creationists therefore assigned to the word "water" a meaning that the biblical author could not have intended. The interpreters, therefore, are not trustworthy scholars—certainly not literalists, as they claim—if they identify liquid water as vapor.

Leading creationists have much to say about God, inspiration, biblical authors, the Bible, and themselves. They say that divine control extended to the actual choice of words by biblical authors— that is, the writers were verbally inspired. They believe, therefore, that authors could transmit messages of a deity through the common idiom in straightforward narratives of historical fact. I am not aware that any creationists have defended the notion that the writer or writers of Genesis used the common term for water but actually meant invisible (molecular) water. That would contradict their notion of perspicuity: that the Bible speaks simply and certainly unambiguously, so that no one should misunderstand.

Biblical writers had no word for the molecular state. Having no visual perception and no mental comprehension of such a state, they could not write about water vapor. And no matter how diligently creationist authors may try to justify their presumptions in developing the vapor story, the biblical authors could write only about visible forms of water.[1] (Light and water relations are taken up in greater detail in chapter 18.)

Layered and Stable Atmospheric Gases. A stable vapor canopy resting upon a supporting layer of heavier gases has been accepted as fact by many creationists, despite a number of arbitrary assumptions. Two observations seem to assure creationists that their idea of a stable vapor canopy raises no serious physical or intellectual problems. Their first observation was that water vapor is "lighter than air."[2] Expanding on this, they recorded that the ratio of the molecular weight of water vapor to that of dry air is 0.622.[3] (That is the correct value relative to the standard atmosphere.) Apparently for creationists this is convincing proof that water vapor should have remained several thousand years as an unmixed layer supported by heavier gases. These authors apparently have little appreciation for Dalton's law of partial pressures. The undeniable implication of that law is that gaseous molecules diffuse independently into space occupied by other gases.

For most canopy models, creationists have nothing to say about incredibly great vapor pressures and densities of water vapor at alleged boundaries between their canopies and the supporting heavier gases. (I give vapor pressure and density estimates for three of their models in chapter 18.)

Light Transmission Problems. Creationist assumptions about light distribution raise many problems. The most basic is the lack of evidence. Creationists have never provided scientific evidence that their massive water canopy could have been maintained above earth and that it was completely transparent to light. I am not aware that any leading creationist has ever considered empirical data that show absorption of light in the red and far-red region of the spectrum. Published absorption values are high enough that a small fraction of the alleged vapor in creationism's massive canopy would have totally blocked transmission in that spectral region and hindered the normal functioning of important plant pigment systems. (I give data and discuss optical properties of atmospheric water in chapter 17.)

Creationists assume that light passed to earth unhindered. Do creationists understand the problems that would have arisen if nearly half of the solar constant (energy in the visible spectrum) could have entered the atmosphere and had been trapped daily for thousands of years under their vapor blanket? (Quantitative values for earth's solar energy budget, heat distribution, and energy balance are given in chapter 14.)

Creationism's Seasonless Planet. Creationist writers seem preoccupied with absorption and retention of solar energy but not with any balancing loss to space. Such loss is a vital factor in earth's present energy regime. Infrared radiation above 0.8 micrometers is the major fraction of the sun's radiant energy arriving at the top of earth's atmosphere. Creationists assume that continuous interception of that longwave energy by their vapor canopy plus the penetrating visible radiation (much of it absorbed and reradiated at longer wavelengths) was evenly distributed over the globe, so that temperatures at all latitudes were virtually identical. (Quantitative data on global energy distributions are given in chapters 14 and 15.)

Creationist leaders can only guess that their imagined world of 4,500 to 10,000 years ago was a seasonless planet. They never describe a mechanism by which their canopy could distribute energy and thus bring a worldwide uniform climate. Having neither rational theory nor empirical evidence, believers are left with specious rhetoric about a mysterious compartment that in some unknown way distributed heat evenly—even at the poles—and, moreover, accomplished this in the absence of winds.

Creationism's Hot and Windless Planet. Creationists assert that the earth was seasonless but also a windless planet. They easily cancel winds of every name and direction, vertical and horizontal, and establish universal doldrums. In imagining such a model, creationists disregard the undeniable fact that when the sun irradiates a hemisphere of earth, it produces uneven heating of surface features and atmosphere. This uneven heating causes localized differences in atmospheric pressure, which in turn account for the universal disturbances we call winds. In the rushing winds we sense directly the effects of a process working to equalize pressure differences.

In imagining a windless planet, creationists must also deny the role of wind-driven ocean currents in maintaining poleward heat distribution and balance needed for a habitable earth. Abolishing the winds, as creationism's model requires, would mean that heat buildup in tropical and subtropical regions could reach lethal temperatures without an effective means of transport to higher latitudes.

Fallacies that creationists have produced concerning their primitive world should be widely exposed. They treat with indifference the primary entities of light, water, heat, gases, and other life-sustaining factors, which also become life-threatening when they are deficient or in excess. Arbitrary manipulations of those factors are expected in a fictional wonderland; they would produce deadly imbalance if imposed on the real world.

Notes

1. Biblical writings contain many words for two visible phases in which water occurs. Scriptures refer to liquid appearing as ordinary water, rain, dew, mist, and clouds. And they speak of the solid phase appearing as frost, hail, ice, and snow. They also identify specific aspects of watery forms (clouds that are thin, thick, heavy, dark), and they speak of clouds and mists vanishing after a brief time (as does human life, see James 4:14). The ancients could read signs (in the Western sky) of approaching weather fronts and showers (Luke 12:54). As for atmospheric water, biblical authors represented their God as making "the clouds rise at the end of the earth" (Psalms 135:7), who also distilled the mist in rain, which the skies drop down abundantly (Job 36: 27, 28). The word "mist" presently serves in modern translations for visible watery particles once called vapor or vapours (see the King James version of 1611). Scriptural words noted above refer to two water phases that people can see. I find no evidence that scriptures provide definitive knowledge of the third phase, the gaseous vapor state we cannot see. Nor am I aware of any physical chemist or other specialist in water science who would attribute such knowledge to biblical authors.

2. J. C. Whitcomb and H. M. Morris, *The Genesis Flood* (Phillipsburg, N.J.: Presbyterian and Reformed Publishing Co., 1961), p. 241.

3. Ibid., p. 256.

13

Water Relationships
Science versus Creationism

Water Distribution: Principles and Problems

The cycling of water between earth and sky is crucial to life. The following factors are involved in that cycle:

1. Free water at surfaces, such as ocean, soils, and plants.

2. Solar heat, the driving force that evaporates water.

3. Winds, from solar heating, that carry moisture from evaporation sites.

4. Atmospheric cooling that condenses water vapor.

5. Condensation nuclei as dust and salt particles (effective but not always necessary).

6. Gravity that causes precipitation of condensed water and the flow of streams and groundwater to oceans or other basins, thereby completing the hydrologic cycle.

Creationist views and those of conventional science generally agree on the first three statements, except that creationists reject strong winds. On the last three, however, there is much controversy, particularly in the way creationists relate the stated factors to their vapor canopy in the alleged preflood era.

Creationists make numerous assumptions about water in developing their flood story. The following statements summarize creationist views about chemical and physical states of the alleged canopy.

1. The vapor was chemically pure, free of volcanic dust and other contaminants.

2. It had great mass, equivalent to much of the present ocean.

3. The vapor was physically stable, unaffected by cooling, density change, or other factors.

4. It was spatially fixed for thousands of years, resting upon and supported by heavier gases.

5. It was very hot, which allegedly helped maintain its physical state above other gases.

6. It was invisible.

7. It was transparent to light.

Creationists infringe upon many sciences when they discuss water. They manipulate oceanography in distributions of ocean water. They implicate meteorology in suppositions about water circulation (or noncirculation) in the atmosphere. They involve geology in assumptions about ocean-bearing crust—which they raise or lower—depending on episodes in the flood story. Extensive caverns of heated and pressurized water also play an important role in creationism's geohydrological conjectures. Obviously, creationists impose strict physical and temporal constraints on water phases.

Water Phase Relationships

Earth's free-water system has three states—liquid, solid, vapor— depending on variables of temperature and pressure. Fortunately, the relations among those variables are precisely defined over a wide range.[1] I have plotted such data to scale as a phase diagram (see fig. 5, solid lines). The figure illustrates a simple function, where a given value of one variable determines the value of the other. This elementary view of phase relations illustrates flaws in the creationist preflood water model.

The starting point is O, which is plotted at 0.46 centimeter mercury pressure (cm Hg) and zero centigrade (0°C). Point O is called the "triple point for water." At that point all three water phases (ice, liquid, vapor) exist at equilibrium. Branching from Point O are four lines or curves: OD, OH, OS, OC.

Line OD, the fusion curve, represents the liquid←→ice (melting←→freezing) equilibrium.[2] The line is not vertical. For every 76 cm Hg pressure increase along line OD, the temperature decreases by 0.0075°C. Creationists believe that water never froze in their preflood world. Therefore, line OD is meaningless. In fact, drastic revision of the figure is needed to represent the creationist model. The revision requires that from a selected "warm" temperature (possibly 15°C above zero), a line must be drawn vertically, and the whole left section of the graph be discarded. The deletion would cut out curve OD as well as OH and OS.

Curve OH is the sublimation curve for ice, plotted only to –90°C. Water molecules at the OH boundary move as two-way traffic, where one lane is the sublimation of ice to vapor. An example of sublimation occurs in cold periods when snow or ice does not melt, but passes directly to the vapor state. Sublimation also occurs in the drying of laundry in a subfreezing wind. Sublimation of an ice-crystal cloud, resulting in its disappearance, is still another example of the ice-to-vapor transition.

Vapor condensation is the opposite process. However, sometimes vapor does not immediately freeze when cooled below zero. Therefore, part of the OH is indefinite. Curve OS gives pressures of supercooled vapor plotted to –40°C. The space between curves OH and OS defines an area of uncertainty in the transition of vapor to the solid phase. Further cooling or increased vapor pressure, or combined effects of these, would be required before the vapor-to-solid transformation would occur.

The area to the left of lines OD and OH represents a complex world of ice and cold, where severe conditions force animals to adapt in various ways or to migrate. For plants, such conditions inevitably account for reduced growth or dieback until dormant parts can regenerate. On the right side of figure 5 is a wholly different world, both in its physical processes and in the behavior of its inhabitants. Although some organisms can tolerate external

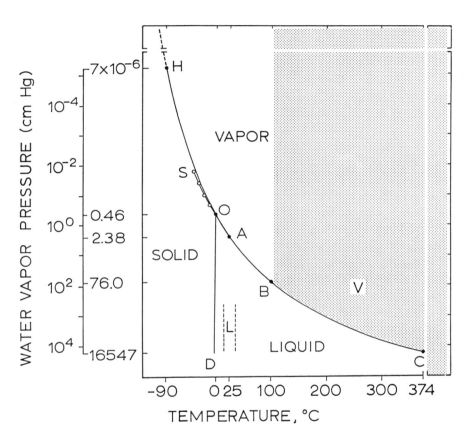

Fig. 5. Water phase equilibria of pure water (solid lines) and an alleged non-equilibrating phase (V), which represents creationism's vapor canopy.

temperatures well below freezing, and others exist near the boiling point of water, the range for healthy functioning of most organisms is from 0 to 38°C (32 to 100°F).

Segment OC extends from 0°C to the critical point for water at 374°C. At that temperature the saturated vapor pressure is more than 16,500 cm Hg (225 kg/cm²).

Water exchange across the liquid⟷vapor equilibrium line (OC) is crucially important for surface life on the globe. Movement of water across that line occurs in the two-lane system. We identify one traffic lane as water evaporation. This occurs when a rise in water temperature, or a lowering of vapor pressure, or a combination of such changes results in release of water vapor to space. The evaporation can take place at a pond's surface, from the internal spongy cells of a leaf, from the surface of a raindrop before it hits the ground, and from innumerable other sites.

Water vapor movement in the other lane occurs when vapor pressure increases, or vapor cools, or a combination of these causes vapor condensation. The condensed vapor may appear, for example, as clouds or fog (low clouds), or dew on a leaf, or as an animal's breath when it strikes cold air.

In the creationist model, not only must the solid phase be eliminated, but a very narrow temperature range must be maintained at earth's surface throughout the alleged preflood era. Vertical broken lines ranging from about 15°C to 35°C (see letter L) represent the universally warm and pleasant world the creationists imagine. Vertical extension of the dashed lines does not represent a specific range of pressure.

The creationists' vapor model allows no precipitation in their preflood era. The baseline for the shaded area, therefore, never contacts curve segment BC. The vertical boundary at 104.4°C accommodates the creationist assumption that vapor existed above boiling. Joseph Dillow assumed that 104.4°C was the temperature at the base of his canopy model.[3] (This will be discussed in chapter 18.) The greatly compressed but unscaled area beyond 374°C allows creationists a wide temperature choice within a superheated vapor range (e.g., up to 1,649°C [3,000°F]) as proposed by leading creationists, Whitcomb and Morris.[4]

At 374°C (705°F), water makes an important transition. Above that temperature water cannot exist as liquid (see fig. 5). Creationists might embrace this fact as proof that their superheated steam blanket was indeed safe from any possibility of condensing and falling to earth. For good measure, perhaps, they invoke the very high temperature (noted above) as a further guarantee of their blanket's integrity as vapor.

At point C, the liquid and vapor phases exist at equilibrium and have the same density. At such extreme heat and pressure the densities of water molecules in both phases would be 0.322 grams per cubic centimeter.[5] One must agree with the understatement by Whitcomb and Morris that delineating in detail the physics and meteorology of their vapor canopy may be difficult.[6] Imagine, if you can, vapor at very high density being supported by and maintained above other gases and, further, that such densities of water vapor could transmit enough light for earth's biological needs.

Creationist Water Phases: Ad Hoc Hydrology

Careful scrutiny of the Morris-Whitcomb preflood model for water distribution must deal with manipulated phases in atmospheric layers and earth compartments. Liquid water is a forbidden phase in creationism's upper atmosphere, and ice is banned everywhere.

By eliminating ice formation, creationists are left with liquid and vapor phases on which they impose rigid constraints. The phases must remain within strict bounds of space, time, and function, according to needs of the evolving story. Their speculations about water phases and reservoirs are illustrated in figure 6 (A to E).

Area A. This upper reservoir, according to creationists, must have persisted as very hot vapor somewhere above earth. Possible locations were presumed to be far in space, in the thermosphere, above the stratosphere, in the upper troposphere, or below 20,000 feet. Creationists can't say exactly where (see fig. 3). They do not allow ocean salt, dust, or any other particles to interfere; the vapor must not condense prematurely. And it must intercept and retain solar energy to keep the whole earth warm. Note that the upper vapor layer did not keep the world moist; that allegedly occurred in atmosphere very close to earth.

Area B. The heavier gases of B extended to earth's surface, and allegedly supported layer A. Layer B remained unmixed with

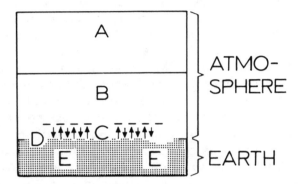

Fig. 6. Creationism's preflood water distribution: alleged atmospheric layers, water phases, compartments. Designations: A. Water vapor blanket. B. Heavier gases extending to earth. C. Aerial vapor/liquid cycle. Arrows depict limited vertical flow and no lateral moisture transport. D. Surface liquid sources (sea, land). E. Cavern liquid sources (suboceanic, continental).

the upper layer for several thousand years. Creationists do not represent this layer as holding water in any form, except in a thin shroud nearest earth.

Area C. Layer C was an inner shell of B that purportedly enveloped earth with high humidity and precipitation.[7] On the subject of rainfall before the flood, Whitcomb and Morris stress two statements about water (Genesis 2:5-6): God had not caused rain on earth, and a mist rose from earth and watered its whole surface. These authors assumed that the no-rainfall regime prevailed from the biblical creation to the Genesis flood. They settle for mist-fall, exclusively in layer C, over the entire planet.

The present hydrologic cycle is strikingly different from the creationist pattern. For example, in the present cycle the average residence time for water in the atmosphere, from the moment of its evaporation to its return to earth's surface, is 9 to 11 days.[8] Flohn further noted, "During this period it is usually carried by the wind over a distance of some 1000 km." For creationists the complete process from evaporation to precipitation took one day. Further, they imagined that the daily cycle occurred over the entire globe and caused "an equable humidity everywhere."[9]

Among the numerous constraints creationists place on their "hydrology" is the no-wind requirement. That assumption would have greatly limited vertical moisture flow. However, they insist that the biblical mist rose, so they allow a small lapse rate and thus a small upward air movement. But their hot canopy hovered above to limit the buoyant rise of water vapor.

Creationists ascribe a universal warming function to their mysterious vapor canopy but do not seriously take up the subject of inversion, which implies temperature increase with height through any intervening layer of air. Why this failure? Is it because they cannot explain how water vapor might cool and approach the dewpoint when rising toward their superheated vapor canopy? Whitcomb and Morris recognized the problem of the hot canopy but also the need for vapor elevation and cooling.[10] Thus, they could only surmise that the lapse rate ". . . was probably small due to the effect of the canopy, so that vapor would tend to recondense and precipitate as a light mist soon after its evaporation."

In creationist speculation, layer C was allegedly a thin aerial layer in which water vapor rose everywhere only a short distance above earth and precipitated daily as a light mist. Precipitation "soon after" evaporation underscores the slight elevation of rising mist and, moreover, that horizontal transport was virtually abolished. (This I have depicted by short vertical arrows—but no lateral ones—in shell C, fig. 6.) One can infer directly, from creationism's no-wind assumption, the limited vertical and horizontal transfer of moisture.

Area D. Reservoir D, in the creationist model, is composed of earth's evaporative surfaces: ocean, lakes, rivers, wet soils, and other sources. On this subject creationists are remarkably imaginative. Whitcomb and Morris invoked Bible verses and from that base developed a fanciful ad hoc hydrology. They see their preflood world as being relatively quiet geologically and with greater land surface than now,

> . . . but the atmosphere was maintained at a comfortable humidity by the low-lying "mist" rising from an intricate network of "seas" (Genesis 1:10) and mildly-flowing "rivers" (Genesis 2:10–14) evidently fed partially or largely by gentle springs.[11]

Notions about rivers and seas have produced a flood of creationist speculation. Consider the following fanciful details:

> On the surface of the primeval world, it is postulated, there was probably an intricate network of narrow seas and waterways whose precise locations need yet to be determined. Though the uniform climate would inhibit air mass movements, as well as storms and heavy rains, a daily cycle of local evaporation and condensation would maintain an equable humidity everywhere.[12]

Fundamentalist authors frequently start with a few key words, embellish them with fanciful rhetoric, and come out with amazing details. Two key words, "seas" and "rivers," are examples. The creation narrative (Genesis 1:9–10) tells of waters brought together into one place and called seas. Creationists promptly take that water, allegedly in a common basin, and spread it out as an "intricate network of 'seas' and mildly-flowing 'rivers.' " From a previous quote we learned that the seas and waterways were "narrow."[13] One can only wonder how ICR creationists find such detailed information. Why were the channels narrow? Does their invented hydrology require that the ocean be spread universally into narrow interconnected waterways? Creationists should be challenged to describe scientifically how their daily water cycle worked to perpetuate humidity and life everywhere—including earth's polar regions—in summer and in the dark months of winter.

In developing stories about water, creationists exploit rivers as readily as they do the sea. They somehow generate information about a global pattern of stream sources, channel distribution, and stream flow; and from these one might conjecture something about watershed and wetlands topography. Are believers to visualize the world—even at polar latitudes—being at one time an expanse of narrow channels, little affected by gravity and "evidently" fed by springs? And if those narrow, interconnected waterways were flowing everywhere on earth, into what body were they finally discharged? Did they flow into themselves? This is circular nonsense.

Fundamentalists believe that they must offer biblical support for what they preach. In this instance they invoke scriptures about rivers (Genesis 2:10–14) and ostensibly find supporting evidence that all rivers in alleged preflood times were languid, mildly flowing streams fed by "gentle springs."

Unquestionably, natural springs associated with regional aquifers are fed by underground water. The water may be slow-moving groundwater from a distant mountain front or more rapidly channeled to the flowing spring. But the flow of groundwater, I emphasize, is gravity-driven, a fact that some creationists don't seem to appreciate. If creationism's rising water mist could not be carried long distances by winds, but fell every day over the globe at local sites of evaporation, they need to answer the following question.

What energy source transferred water to watersheds at higher elevations (higher than ICR's postulated "gentle springs") to begin the downward flow and eventual recharge of their imaginary springs? It appears that creationists have achieved a perpetual motion that transfers water in a definite cycle without a verifiable source of energy.

I question again the behavior of creationist writers who invoke biblical writings in discussing scientific matters. How do creationists who cite biblical authority reconcile their languid flow of rivers with the etymologies and meanings of the river's names?

Expositions on the names and historical meaning of Genesis rivers (Hiddekel [Tigris], Gihon, Euphrates) include such terms as "lively," "bursting," "rapid," and "free-flowing."[14] Such words indicate flow patterns very different from the suppressed and sluggish movement imagined by ICR authors.

If Young's analysis is correct, then creationist leaders have not carefully followed methods expected of scientific scholars. Rather, they have appealed to scripture and abused words to extract the meanings they desired. Scientists, I believe, will conclude that creationist authors have no reliable evidence for making sluggishness the pattern typical of streams everywhere on earth. Nor have those authors given evidence that the physiography of earth would allow a universally tranquil flow of streams in any period of history. Without that evidence, their statements about water distribution noted above remain a complete fabrication.

Area E. The alleged underground reservoirs are no more valid than a mysterious vapor blanket or a rising mist that daily watered earth's total land surface. Yet, creationists, in the absence of scientific evidence, apparently feel secure in conjecture and in challenging opponents to prove that their mysterious water sources did not exist. The beginnings of the earth-cavern and hot water story rest on such biblical words or combinations as "fountains," "deep," "the deep," "fountains of the great deep," and "fountains . . . broken up."

The following is a synthesis on "the deep" as understood by leading creationists. The statements are based mostly on writings from Whitcomb and Morris.[15]

"The deep" or "great deep" is to be understood as liquid water that first covered earth, or the fraction of that water later "segregated" above earth,[16] or as ocean depths,[17] or as hot pressurized water trapped in earth's crust.[18] The "deep" obviously is another mysterious compartment that can serve creationists in manipulations of water in any direction and for numerous functions.

Water allegedly trapped underground at creation built up pressure and temperature steadily[19] until its explosive escape "through crustal fractures all around the earth when the fountains of the great deep were broken up."[20]

The breakup of the fountains of the deep refers to ". . . ocean basins . . . fractured and uplifted sufficiently to pour water over the continents."[21] Whitcomb asserted that the breakup should be understood as "the uplift of ocean floors."[22]

Water and certain reserves of water were created for life and, also, for destruction and death. Morris and the ICR staff were explicit:

> The primeval creation of those two vast bodies of water, one above the troposphere and the other deep in the earth's crust, would thus serve the dual purpose of providing a perfect environment for terrestrial life and also for transmitting the energy for the universal cataclysm which later would destroy that life.[23]

These stated beliefs raise serious concerns. They involve problems of identity, function, and "purpose" within a matrix of surmise, ambiguity, and contradiction. How, for example, can one visualize the alleged fountain breakup as fluids gushing for months from "crustal fractures all over the earth" and, at the same time, as representing specifically "the uplift of ocean floors"? Further, how could raised ocean floors, fractured throughout, serve as an ocean-bearing platform and hold much of earth's 326 million cubic miles of water at a sea level above earth's highest mountains for months before collapsing?

Creationism's "dual purpose" for created bodies of water sounds a teleological note. Their God had plans for animal extinction. In a similar context, Whitcomb and Morris speculated that extinct animals, such as the dinosaurs, could have been on Noah's ark (as "very young animals") but died out after the flood.[24] But they concluded otherwise: ". . . it seems more likely, however, that animals of this sort were not taken on the ark at all, for the very reason of their intended extinction." Deity had created water reserves for a definite killing time.

Notes

1. H. Flohn, *Climate and Weather* (New York: McGraw-Hill, 1969), pp. 44, 45; R. C. Weast (ed.), *CRC Handbook of Chemistry and Physics,* 65th ed. (Boca Raton, Fla.: CRC Press, 1984), pp. D192–94.

2. For liquid $\leftrightarrow$ ice equilibrium, the number of water molecules freezing across a unit surface area per unit time equals the number melting from that surface.

3. J. C. Dillow, *The Waters Above: Earth's Pre-Flood Vapor Canopy* (Chicago: Moody Press, 1982).

4. J. C. Whitcomb and H. M. Morris, *The Genesis Flood* (Phillipsburg, N.J.: Presbyterian and Reformed Publishing Co., 1961), p. 240.

5. L. Haar, J. S. Gallagher, and G. S. Kell, *NBS/NRC Steam Tables: Thermodynamic and Transport Properties and Computer Programs for Vapor and Liquid States of Water in SI Units* (New York: Hemisphere Publishing Corporation, 1984), p. 15.

6. Whitcomb and Morris, *The Genesis Flood,* p. 240.

7. Ibid., pp. 241–43.

8. Flohn, *Climate and Weather,* p. 36.

9. H. M. Morris, *Scientific Creationism,* General Edition (El Cajon, Calif.: Master Books, 1974), p. 125.

10. Whitcomb and Morris, *The Genesis Flood,* p. 242.

11. Ibid., p. 243.

12. Morris, *Scientific Creationism,* p. 125.

13. Ibid.

14. R. Young, *Analytical Concordance to the Bible* (Grand Rapids, Mich.: Eerdmans Publishing Co., 1952), pp. 309, 479, 753.

15. Whitcomb and Morris, *The Genesis Flood.*

16. Ibid., pp. 9, 42.

17. Ibid., p. 9.

18. Ibid., p. 242.

19. Ibid.

20. Ibid., p. 261.

21. Ibid., p. 9.

22. J. C. Whitcomb, *The World That Perished* (Grand Rapids, Mich.: Baker Book House, 1973), p. 35.

23. Morris, *Scientific Creationism,* p. 125.

24. Whitcomb and Morris, *The Genesis Flood,* p. 69n.

14

Solar Energy and Earth

Energy Science: Quantity Measurements and Numbers

The words of William Thomson (Lord Kelvin) set a standard for those who would be scientists:

> I often say that when you can measure what you are speaking about, and express it in numbers, you know something about it; but when you cannot express it in numbers, your knowledge is of a meagre and unsatisfactory kind; it may be the beginning of knowledge, but you have scarcely, in your thoughts, advanced to the stage of Science, whatever the matter may be.[1]

In this chapter I present scientific data for earth's annual solar energy income, flow, and balancing losses to space. I then compare the data with ideas advanced by creationists.

Light and heat are essentials for life, and such energy flows continuously from the sun. Except for a very small amount that is temporarily fixed, the annual energy loss to space balances the income. Without that balancing loss, global temperature would rise and threaten life on earth.

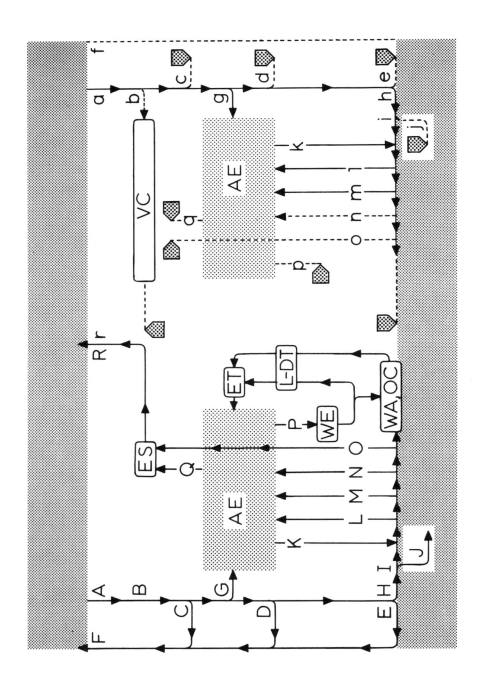

Fig. 7. Annual solar energy flow within earth's surface media. Contrasted are distribution, loss, and resulting energy balance or imbalance, for two models: science and creationism. Values in brackets are to be multiplied by 10^{22} joules per year.

(A)   and (a) Incident radiation for each model [550].

(B)   Absence of a universal cloudless canopy. (b) Alleged vapor canopy (VC) absorbs all ultraviolet (UV) and infrared (IR) radiation [308], but transmits visible spectra.

(C)   Shortwave reflection by clouds [127]. (c) No VC clouds, therefore no reflection.

(D)   Light back-scattering [28]. (d) No back-scatter published; emphasis on absorption and retention.

(E)   Surface reflection [17]. (e) No reflection value published.

(F)   Albedo, sum of C + D + E [172]. (f) No albedo published.

(G)   Atmospheric absorption [110]. (g) Absorption estimate only by surmise (see h).

(H)   Surface absorption and heating [268]. (h) Surface absorption and heating [242]. The H and h values include scattered, diffuse, and direct radiation.

The following categories (I to R) relate strictly to energy transformations and transfers of the G + H energy pool.

(I)   Photosynthetically fixed energy [0.29]. (i) Assumed universal plant production; no energy estimate given.

(J)   Bio-derived energy stored as fuel [10^{-5} to 10^{-4}]. (j) No preflood fossil fuels produced.

(K)   IR radiation from atmosphere to surface matter [578]. (k) No data available.

(L)   IR radiation from earth to absorbing atmosphere [555]. (l) No data available.

(M)   Latent heat transported to atmosphere by water vapor [126]. (m) Heat transport assumed; no data available.

(N)   Heat transported by convection or turbulence to atmosphere [55]. (n) General rejection of turbulence.

(O)   IR emission directly from earth through atmosphere to space [110]. (o) No IR emission to space.

(P)   Energy distribution loop. Wind energy (WE) generates wave action (WA), drives ocean currents (OC), and carries heat and moisture globally in long-distance transport (L-DT). (Energy quantities and transfers [ET] are discussed in the text.) (p) Denial of winds cancels all wind-related energy flow.

(Q)   IR from atmosphere emitted to space (ES) [268]. (q) No release of IR radiation. Heat retained under the dominant vapor canopy.

(R)   Total IR loss to space [O + Q = 378] balances G + H value. (r) No confirming evidence available for IR release.

Energy Budget and Balance

The quantity of energy that reaches earth's atmosphere from the sun can be estimated from values for earth's surface area, the solar constant for radiation, the time of exposure, and the surface effectively exposed. This last geometric factor is not intuitively obvious.

The revolving earth has one hemisphere in darkness while the other is exposed—much of it obliquely—to the sun. Earth's spherical geometry thus determines surface orientation and its exposure to the solar energy.

The surface of earth effectively exposed, relative to its total surface, has the value $\pi r^2 / 4 \pi r^2$, which reduces to ¼ or 0.25. Thus we can estimate solar energy arriving annually at the top of earth's atmosphere by multiplying the factors (0.25) $(5.1 \times 10^{14}$ m^2 [the earth's total surface]) $(1370$ J/m$^2 \cdot$ s [the solar constant]) $(3.15 \times 10^7$ seconds/year). The answer is 5.5×10^{24} joules per year. This (expressed as 550×10^{22} J/yr) is the value given for symbols A and a in figure 7.

Figure 7 appears to be a formidable maze of lettered paths and arrows. The overall view shows two models side by side (scientific and creationist), identified, respectively, by capital and lower case letters. Solid lines represent energy flow; broken lines indicate questionable flow. Large hatched arrows indicate total blockage or restricted flow of energy. Assignments of energy quantities for lettered paths are in the figure legend.

The following brief sketch will give readers a preliminary view of energy transfers. Energy enters at A and moves along diverging courses. Paths C, D, E, and F represent shortwave (light) energy quantities reflected or scattered back to space; they do not enter earth's functional energy account. Creationists have been preoccupied with retention and distribution of solar energy under their vapor canopy and, apparently, have not recognized a need for diversion of energy (c, d, e, f) to outer space.

Energy that arrives at A (and avoids the direct shunt to space) moves to atmosphere (G) and to the ground (H) where it provides warmth, causes photosynthesis (I), and eventually contributes very small amounts as fossil fuel (J). Fossils [j] allegedly could not occur in creationism's preflood times.

Energy moves along several paths between earth and

atmosphere (K, L, M, N). Some energy escapes from earth's surface directly to space (O), but larger amounts are released from the upper atmosphere (Q). The o and q paths are blocked by creationism's vapor canopy.

Another important difference between the models is the energy transport loop (P) caused by winds. Creationists deny the existence of significant winds (p).

Model Comparisons: Energy Flow and Balance

Energy quantities A to R (fig. 7, legend) represent a consensus attained over many years by scientists in the fields of solar radiation and the energetics of earth and atmosphere.[2] The solar constant used in the earlier computation, 1,370 joules/m²·s (or 1.96 cal/cm² · min), derives from 1,370 watts/m², an averaged value from satellite and other data.[3] Hereafter, I will refer to numerical energy coefficients (fractions of 550) but not always append the constant, 10^{22} J/yr.

To simplify comparisons between science and creationist views, I have divided figure 7 into nine parts to discuss groups of related subjects. The purpose is to show contrast, and possible agreement, by discussing topics side by side (e.g., B versus b, C versus c, and so on).

Incoming Energy. Perhaps some creationists will agree that coefficient 550 (as qualified above) reflects a reasonable annual solar income for their preflood period. Creationists credit the worldwide warmth of that period to heat retention by water vapor in the atmosphere and ". . . *not* due to an actual increase in radiation from the sun . . ." (emphasis in original).[4] The value 308 at symbol b accounts for about 56 percent of solar spectral energy arriving annually at earth's upper atmosphere and represents ultraviolet and infrared radiation allegedly absorbed by creationism's vapor canopy. No vaporous water blanket (b) is known to have enveloped earth at any time.

If all present atmospheric water existed as a liquid shell, it would envelop earth to a depth of about 2.5 cm (1 inch). However, atmospheric water disperses above earth at greatly varying heights and densities, and it exists either as *vapor* or as *clouds* of liquid or ice crystals. Indeed, that relatively small amount of water mediates important energy transfers between atmosphere and earth.

Losses to Space. Some shortwave (visible) solar radiation is not involved in earth's energy budget (C, D, E, with F being the sum of the annual losses to space). Creationists declare that no clouds ever formed in their preflood sky; thus, there could have been no reflection (c) from clouds. I have not found creationist estimates for light back-scattering to space (d) or reflection from earth's surface matter (e) in the alleged preflood era. Therefore, in the creationist model, the earth must accommodate an additional one-third of the present annual solar income. Creationists need to answer several questions. How could their windless planet distribute the excess heat? And how could life in their warm, shallow seas or on land have endured the cumulative heat burden?

Energy Flow. We started with a total solar irradiance of 550 (times our understood constant of 10^{22} J/yr). We lost 172 as shortwave energy shunted back to space. This left a net income of 378 distributed two ways: to atmosphere (G) and to surface (H).

In contrast, the creationist model is complicated by unverified surmise. Starting with an annual solar income presumed to be around 550, creationists immediately trap more than 300 (all IR and UV energy) in their vapor blanket. This leaves about 44 percent (via the visible spectrum) to be apportioned between atmospheric absorption (g) and surface absorption (h). I have assigned no value for energy absorbed by atmosphere. Whatever one's guess might be, the alleged absorption based on their model would have taken place below their alleged vapor canopy. I am not aware that any creationist has ever suggested quantities of energy absorbed by, or energy transfers between, atmosphere and earth.

The above discussions have involved general aspects of energy flow. From this point, my discussion will focus on various distributions and forms of energy: fixed energy in plants, bound energy in fuels, latent energy in evaporated water, sensible energy in heated atmosphere, and kinetic energy in winds.

Stored Energy. The estimated energy stored annually by photosynthesis seems small and insignificant (I = 0.29; the ratio 0.29/550 = 1/1897 or 0.00053).[5] Nevertheless, that fraction is at the base of the food web from which herbivores and, indirectly, carnivores as well as omnivores extract chemical energy for life.

Although creationists give no estimate for annual plant production through photosynthesis (i), they conjecture that vegetation produced was much larger before the flood than at present because of universally favorable soil fertility, mist irrigation, and climatic conditions. They assert that biomass stripped from the ground and buried by a universal flood accounts for earth's present coal deposits.[6]

The estimate by Smith for the annual energy contribution of biomass to fossil fuels (J) is 10^{17} to 10^{18} J/yr.[7] Creationists deny that fossil fuels accumulated through geological ages. They believe that burial of organic matter from which fuels were produced occurred in their flood year.[8]

Atmospheric Energy. Several paths in figure 7 connect earth's surface with a compartment I call atmospheric energy (AE). Path K depicts infrared (heat) radiated from that energy pool to earth. Obviously, the energy assigned at K (578) is larger than the absorbed energy (110) at G. It is also larger than G + H combined (378) and exceeds even the total irradiance (550) at A. The reason is vertical cycling as energy is transformed and exchanged globally between atmosphere and earth.

Heat exchange on path L involves IR radiation (555) from earth to atmosphere. And the vertical path, M, depicts transport of a latent heat quantity (126). Such energy is bound in water molecules when they evaporate and is released as sensible heat when they condense.

It is not possible to assign quantities for the earth-atmosphere energy transfers (k and l) in the creationist model. Although such pathways might be depicted for energy flow, no available data confirm such flow.

Energy Transfer. Leading creationists in ICR offer unverified speculations for latent heat that rises from earth in evaporated water (path m). They introduce the postulate that a global system of interconnected waterways accounted for a "daily cycle of local evaporation and precipitation" that maintained "equable humidity everywhere."[9] They make no attempt to compute energy quantities transported.

Path n suggests transport of sensible energy to the atmosphere. But creationists deny any strong updrafts of heated air. Having abolished wind in their preflood period, creationists cannot accept turbulence or any strong convective movement of heat to the atmosphere. Consistent with this constraint is their assumption that water vapor rose everywhere only a short distance before it condensed as a light mist. They do not allow winds to carry moisture long distances, either vertically or laterally, in the alleged preflood period.

Radiated Energy. Solar irradiance at earth's surface (H) heats rocks, soils, and so forth, which eventually radiate IR energy to the environment, especially to atmosphere (path L). The IR not absorbed by clouds, gases, or other matter readily escapes to space. Those emissions to space often take place at night under conditions of low humidity and clear skies. Such heat losses have been called nocturnal or radiational cooling.

Comparisons of path O versus o show striking differences. Creationists give no estimates of surface heat radiated from their preflood earth. Actually, IR radiation to space (path o) would have been impossible if their basic assumptions were true. If the vapor canopy (VC) absorbed all incoming IR radiation (path b), as creationists affirm, it should also have blocked the outgoing IR radiation.

Wind Energy. Path P, representing energy transported by wind, can be visualized as a loop that arises in and returns to the atmosphere. Wind arises from uneven heating of atmosphere. The energy of wind, variously dissipated and transformed, returns to the atmosphere as heat. In the figure, wind energy (WE) transfers kinetic energy to surface water, causing wave action (WA) and ocean currents (OC). Such currents may run shallow or deep, depending on the strength of prevailing winds.[10] Wind and ocean currents are crucial in moderating distant weather and climate as they carry tropical heat toward and into polar regions.

Smith estimated that the total kinetic energy of winds over the planet is equivalent to about 10 percent of the absorbed solar energy.[11] This would be about 38×10^{22} J/yr (i.e., 10 percent of the G + H energy value [fig. 7]).

In addition to wind-driven currents of tropically heated water, the major mechanism for transporting surplus energy poleward are large traveling systems of circulating winds (cyclones and anticyclones). These mix moist air from the tropics with dry air from polar regions.

Creationists invent a very different scenario. They abolish wind throughout their preflood period (path p, fig. 7, is blocked). They allow no waves or ocean currents, no traveling eddies of low and high pressures in a global circulation—no variable weather, no clouds in the sky, no seasons. The sun shone and mist fell every day, and pleasant warmth was everywhere, even at the poles. All this took place, creationists fantasize, in a marvelous greenhouse world, under a mysterious water canopy.

Heat Released to Space. The atmosphere releases IR heat to space (path Q) after receiving large amounts of energy annually from earth (paths G, L, M, N). The creationist model cannot allow IR to radiate from atmosphere to space via path q, for the same reason as that given above for path o. Creationism's massive and very hot vapor blanket above the atmosphere of heavier gases would have blocked IR escape. The vaporous blanket, without heat controls, could have driven temperatures above tolerable heat limits of the biosphere.

Annual Energy Bookkeeping

Over the past half century, atmospheric and earth scientists have developed a useful system of energy accounting. They have monitored and classified electromagnetic energy spectra and determined rates and quantities of flow. In summarizing accounts of energy transfers, they have consistently emphasized that a delicate balance must be maintained between incoming and outgoing radiation; otherwise, earth's surface temperatures could become lethal.

To illustrate the annual global balance, I have compiled in table 1 values from figure 7 (legend). Columns at the left show shortwave energy that enters but is not retained in earth's environment. Scattered and reflected, the quanta escape to space at the speed of light. The other energy categories (space, surface, atmosphere) also show equal in!-!out distributions.[12]

Energy accounts obviously cannot balance where energy-absorbing substances accumulate in the atmosphere and hold increasing amounts of heat—a present concern in studies of the so-called greenhouse effect.

Much attention now focuses on the harmful effects that relatively small increases of atmospheric gases have on world ecology. In contrast, the creationists clutch their mysterious vapor blanket that allegedly held millions of cubic miles of water. To their blanket they ascribe marvelous powers: it absorbed and retained and distributed heat; it abolished seasons; and purportedly it safeguarded life and promoted abundant growth everywhere.

Table 1. Summary of annual global heat balance in categories of space, surface, and atmosphere, plus transits of nonreactive shortwave energy. See lettered paths identified in figure 7 and the legend. Column totals at *. Multiply values × 10^{22} J/yr.

Nonreactive Shortwave Balance		Reactive Energy Balance					
Transits		Space		Surface		Atmosphere	
In	Out	In	Out	In	Out	In	Out
C 127	127	G 110	O 110	H 268	L 555	G 110	K 578
D 28	28	H 268	Q 268	K 578	M 126	L 555	Q 268
E 17	17				N 55	M 126	
					O 110	N 55	
* 172	172	378	378	846	846	846	846

Creationists make sweeping and arbitrary statements about solar energy and earth. They have no quantified measurements. Their energy books don't balance.

Notes

1. D. E. Tilley, *Contemporary College Physics* (Menlo Park, Calif.: Benjamin/ Cummings Publishing Company, 1979), p. 284.

2. In most of the sources named below are tables and diagrams that summarize data on solar irradiance and the heat balance of earth and its atmosphere. H. G. Houghton, "On the Annual Heat Balance of the Northern Hemisphere," *Journal of Meteorology* 11, no. 1 (February 1954): 1-9; J. C. Johnson, *Physical Meteorology* (Cambridge, Mass.: The MIT Press, 1954); H. Flohn, *Climate and Weather* (New York: McGraw-Hill, 1969); K. Ya Kondratyev, *Radiation in the Atmosphere,* International Geophysics Series, vol. 12 (New York: Academic Press, 1969); H. R. Byers, *General Meteorology,* 4th ed. (New York: McGraw-Hill, 1974); J. Ellis and T. H. Vonder Haar, "Zonal average radiation budget measurements from satellites for climate studies," *Atmospheric Science Paper No. 240* (Fort Collins, Colo.: Colorado State University, 1976); D. M. Gates, *Biophysical Ecology* (New York: Springer-Verlag, 1980); R. Anthes, J. J. Cahir, A. B. Fraser, and H. A. Panofsky, *The Atmosphere* (Columbus, Ohio: Charles E. Merrill Co., 1981); H. G. Houghton, *Physical Meteorology* (Cambridge, Mass.: The MIT Press, 1985); L. D. Smith, T. H. Vonder Haar, and D. L. Randel,

"Interannual Variability Study of the Earth Radiation Budget from Nimbus 7 Monthly Data" (AMS Sixth Conference on Atmospheric Radiation/Satellite Meteorology Joint Session, Williamsburg, Virginia, May 1986) (Fort Collins, Colo.: Department of Atmospheric Science, Colorado State University, 1986); D. G. Smith (ed.), *The Cambridge Encyclopedia of Earth Sciences* (New York: Crown Publishers/Cambridge University Press, 1981); R. C. Weast, *CRC Handbook of Chemistry and Physics* (Boca Raton, Fla.: CRC Press, 1984).

3. Houghton, *Physical Meteorology*, pp. 71, 73.

4. J. C. Whitcomb and H. M. Morris, *The Genesis Flood* (Phillipsburg, N.J.: Presbyterian and Reformed Publishing Co., 1961), p. 255.

5. P. S. Nobel, *Biophysical Plant Physiology and Ecology* (New York: W. H. Freeman, 1983), p. 332.

6. Whitcomb and Morris, *The Genesis Flood*, pp. 277–79.

7. Smith, *The Cambridge Encyclopedia of Earth Sciences*, p. 182.

8. Whitcomb and Morris, *The Genesis Flood*, pp. 434–37.

9. H. M. Morris, *Scientific Creationism*, General edition (El Cajon, Calif.: Master Books, 1974, p. 125.

10. P. P. Niiler and C. J. Koblinsky, "A Local Time-Dependent Sverdrup Balance in the Eastern North Pacific Ocean," *Science* 229 (1985): 754–56; S. Weisburd, "Deep-sea Currents Driven by Winds," *Science News* 128, no. 9 (1985): 141.

11. Smith, *The Cambridge Encyclopedia of Earth Sciences*, p. 143.

12. H. R. Byers (*General Meteorology*, 4th ed. [New York: McGraw-Hill, 1974], p. 50) has published a similar table in which he presents data slightly modified from H. G. Houghton, "On the Annual Heat Balance of the Northern Hemisphere," *Journal of Meteorology* (Cambridge, Mass.: The MIT Press, 1954), p. 8.

15

Earth Satellite Energy Profiles versus Creationism

Earth Orbiter Measurements: Absorption and Emission

As discussed in chapter 14, nearly 70 percent of the solar irradiance is absorbed by atmosphere and earth. That percentage, however, tells nothing about the differences in amounts absorbed per unit area at various latitudes. Those differing quantities are important in understanding earth's energy transport patterns and heat balance and will be useful in examining creationist ideas about global energy distribution. Creationists generally assume that solar energy was distributed over the globe by water vapor—without involving winds and ocean currents in the transfer process.

Earlier I discussed annual infrared radiations to space, but did not disclose the differing quantities of IR radiated from defined geographic belts. Satellites collect those data and thereby increase our knowledge of the global dynamics of energy transfer.

Ellis and Vonder Haar[1] published large-scale collections of satellite measurements along with statistical and graphical data on energy transfers. Plotted in figure 8 are two columns of their data.[2] Energy was measured in 10-degree zones; for example, zone 2 at 15°N denotes averaged energy (watts absorbed and emitted) per square meter in the globe-encircling belt, 10°N–20°N.

Two energy categories are compared in figure 8: solar radiation that the earth and atmosphere absorbed (A) and the longwave

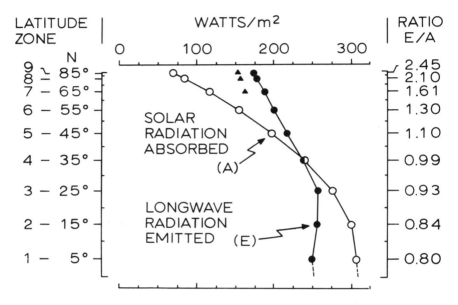

Fig. 8. Solar energy absorbed and infrared energy radiated annually to space by earth and atmosphere. Curves present zonally averaged satellite data. Each point gives absorbed or emitted watts, averaged over 12 months, for each square meter in each ten-degree latitude zone around the earth. For comparison, inserted triangles show average January heat emissions to space from zones 7, 8, 9. Satellite readings for those zones in other months are in figure 9. Data are from J. S. Ellis and T. H. Vonder Haar, "Zonal average radiation budget measurements from satellites for climate studies," *Atmospheric Science Paper* no. 240 (Fort Collins, Colo.: Colorado State University, 1976), p. 48.

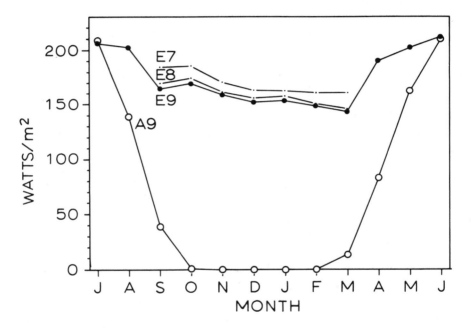

Fig. 9. Zonally averaged solar energy absorbed (A) and thermal energy released to space (E). As in figure 8, zone 7 refers to north latitude 60°-70°, zone 8 is 70°-80°, and zone 9 is 80°-90°. Curves give satellite data from J. S. Ellis and T. H. Vonder Haar, "Zonal average radiation budget measurements from satellites for climate studies," *Atmospheric Science Paper* no. 240 (Fort Collins, Colo.: Colorado State University, 1976), pp. 16–38.

IR energy they emitted (E). The curves indicate a steady decrease in the energy absorbed between the equator and the pole.

Winter brings darkness to one pole or the other for several months with no possibility of solar energy absorption. However, considerable quantities of energy, plotted on curve A, seem to have been absorbed in the three most northerly zones (7, 8, 9) during winter. This is an artifact of statistical averaging, where totals absorbed in months that received sunlight were spread over the year.

The radiation curve for emission is remarkably different. In the tropics, losses to space remained below quantities absorbed (E/A was less than 1). But from the crossover point,[3] curve E diverges until E/A ratios near the pole were larger than two. Most of the region from 60°N to 90°N is within the Arctic Circle, where between 1.5 and 2.5 times the energy absorbed is radiated to space. Those annual averages, of course, cover the dark months, when absorption remains at zero.

The three triangles on figure 8 show energy quantities radiated to space in January, when three zones were in total darkness. Quite strikingly, quantities of heat radiated from the zones in midwinter were only slightly below the annual average shown on curve E. Figure 9 gives radiation values from the same zones, but the averages cover twelve months, July to June.

Satellite emission readings for winter months illustrate that throughout the polar darkness, thermal energy continues to radiate to space at relatively high rates (fig. 9). Emissions from latitude belt 9 (curve E9) show that outgoing heat for October was 169.0 W/m². Later measurements in February showed heat loss of 148.2 W/m², a decrease of only 12.3 percent from the October reading. Corresponding decreases in the other latitude zones for the same months were 11.9 percent for E8 and 12.9 percent for E7.

Meteorological evidence demonstrates clearly that global winds and ocean currents sustain the release of heat from polar regions to space. Creationist leaders, on the other hand, have rejected the role of winds and ocean currents, and have assigned global heat distribution and balance to a water vapor canopy. They do not account for any balancing heat loss to space in their preflood world.

The watts/m² absorbed from March to September (curve A9, fig. 9) totaled 852. The sum of watts emitted in that period from zone E9 was 1,315. Thus, in seven months the energy released to space was 1.5 times the radiation absorbed. For the five winter months, October to February, the absorption was almost zero (only 0.5 W/m² in October), but the total energy radiated to space was 781. The value 781 compared to 852 tells us that radiation to space in five winter months was only 8 percent less than absorption in the seven months extending through the summer. The general flow of energy into and out of Antarctica (not shown here) was almost identical to that of the Arctic.

The satellite data given above demonstrate that in northern latitudes, as elsewhere, dynamic processes operate in an open system. From March to October, Arctic insolation is largely reflected back to space. Most of the energy that does flow into the system (and is later radiated to space) arrives continuously in winter and summer as heat imported from the tropics.

Not only does insolation vary seasonally, but earth's surface materials absorb solar energy at different rates and have different capacities. The reservoirs of atmosphere and ocean are able to absorb large amounts of heat. Moreover, their fluid movement as winds and currents become global forces in the transport of heat.

Creationist Views: Earth, Sun, and Energy Imbalance

Previous chapters have discussed physical entities (earth, sun, air, land, water) and introduced descriptive concepts relating to those entities (heliocentric orbit, light, radiation exposure, energy absorption, heating, heat transfer, seasons, winds, currents). One might expect thoughtful creationists to agree that those concepts truly represent present conditions on earth. Most biblical literalists, however, will assert that certain laws and dynamic processes which presently operate on the orbiting earth were not in force during the biblical creation and flood. Creationists have their own set of rules and explanations.

Earth and Natural Laws. The early earth had light, water, land, and vegetation, but did not orbit the sun. The creation story (Genesis 1:1–19) states that at first the earth existed in darkness. Light (not sunlight) was then created, water was collected into seas, dry land appeared, and seed-bearing plants were established—all of this before the sun and other stars were created.

Scientists cannot accept the upside-down sequence of events that biblical literalists embrace. Theistic evolutionists (who believe that God put evolutionary forces to work) also raise intellectual difficulties. Although they believe that creation days in the Genesis narrative represent geological ages, they still must account for an inverted sequence of origins. They need to explain the origin of plants before sunlight, and must tell how plants evolved over unspecified millions of years on a globe that did not orbit a light-giving star. Would creationists think the suggestion heretical that earth must have orbited something?

Understandably, biblical literalists who give primacy to earth as man's specially designed home might expect sun and stars to have been created after earth and also that they would orbit the earth. Bible verses (Joshua 10:12–13) record that by command from Joshua "the sun stood still" about a day over a specific site, Gibeon, and the moon also "in the valley of Aijalon" while Israel routed the enemy. Then, presumably, earth's stationary satellites resumed their geocentric orbits.

Creationists at the ICR cling to the biblical order of events and, therefore, agree that the sun could not have illuminated earth before the fourth creation day.[4] Thus, it seemed necessary that they postulate another mysterious force, "the primeval light," which, they explain, "For practical purposes . . . must essentially have come from the same directions as it would later when the permanent light-sources were set in place." After advising readers about essential "directions" of that incoming light, these creationists apparently declined to enlighten readers further on the "practical purposes" or to suggest what the earth might have been orbiting in that alleged pre-sun time.

Throughout the Bible from Genesis to Revelation, writers violate the constancy of natural laws, and many creationists subscribe to those violations. They believe the biblical account that earth existed before the sun (Genesis 1:1, 16–18) and that someday stars will fall to earth (Revelation 6:13–14). Do modern creationists believe literally that a time will come when a great quake shakes the earth, the sun goes black, and stars fall to earth like a "fig tree [that] sheds its winter fruit when shaken by a gale"? It seems likely that John saw the incandescence of small particles ("shooting stars") falling to earth and imagined that one day all the stars of heaven would fall to earth.

Those who appreciate the dynamics of orbiting bodies (gravitational astronomy) understand that stars are not the size of figs. They do not fall to earth. Religious fundamentalists apparently often misperceive natural relations. To them, natural laws overturned and presented upside down have special appeal and usefulness.

Creationism's Preflood World and Energy Flow. The alleged preflood world was an energy-receiving system: a uniformly warm, moist, windless, seasonless biosphere. After the sun was created, earth received solar energy, but none was lost back to space. Although this heat loss is important in earth's heat balance, creationists have given it little attention.

Creationists have built their unique and complex story around a massive imaginary shell of water vapor. Henry Morris conjectured, "With the antediluvian vapor canopy, the earth must have enjoyed a uniform, mildly warm climate all year long and over its whole surface."[5] Seasons did not exist, even at the poles. Allegedly the canopy canceled seasonal effects that naturally result from earth's inclined axis and orbit around the sun. Faithful believers are never told specifically how the canopy accomplished this moderating effect. Whitcomb and Morris seemed to apologize for their lack of such knowledge:

> The physics and meteorology of such a vapor canopy, and its maintenance in the antediluvian atmosphere, may be difficult to delineate in detail; even today little is known about the *present* atmosphere, its constituents and physical behavior.[6] (Emphasis in original)

Not surprisingly, these authors have never provided scientific details (physical or meteorological) about their vapor canopy. That shortcoming, however, has not hindered the continuing development of their vapor story.

Apparently, one has only to believe that the vapor canopy regulated temperature universally by absorbing and spreading solar energy evenly over the globe. In that uniform distribution, the canopy somehow eliminated differential rates of radiation exposure on the spherical earth. Satellite measurements have shown that solar irradiance directed on a sphere decreases steadily with increasing distance from the vertical source. Creationists should explain scientifically how solar energy was distributed evenly over the globe and seasons were abolished on their pre-flood earth.

Did the canopy also abolish differential heat absorption, regardless of differences in absorption rates of earth's surface matter? Moreover, the canopy had to prevent massive air movements and suppress all local windstorm activity and associated fronts. Creationists expect people to believe all this—but apparently never attain a solid base of scientific evidence.

Creationism's Heat Machine: A Potential Hazard

Creationists have taken solar energy and earth and have fashioned an implausible heat engine. Undeniably, they had a constant heat source—the solar furnace. But creationists had problems with engine design. Their boiling vapor canopy would have been too dense to let light through, in which case the engine would be useless. Creationists deny this; their vapor blanket, they presumed, was transparent. If that were true, creationists would face another problem. Their heat-absorbing canopy had no safety outlets and thus no way to release excess heat from the system.

What are the terrestrial and atmospheric mechanisms that allow visible and longwave energy to be returned to space and thereby release excess heat outside the system? Three processes can be defined as escape mechanisms.

The first is *albedo,* that amount of energy reflected back to space. As noted in chapter 14, the average albedo recorded in seven years of Nimbus-7 orbits was 31.3 percent of the annual incoming energy. What happened to that large quantity of radiation in the creationists' world? Albedo, backscatter, and reflection are words apparently not in the creationist vocabulary.

The second mechanism that allows heat to escape is direct infrared radiation from earth to space. This often takes place on cloudless nights or through a low-humidity atmosphere. Annually such emissions amount to about 20 percent of the total incoming energy. In the way creationists describe their vapor canopy, it would not have allowed infrared energy to escape to space.

The third mechanism for heat escape is the infrared radiation from the atmosphere, mostly from the upper troposphere. Annually this is nearly half of the total incoming solar radiation. The infrared emission is long wavelength radiation, which creationists have decided could not pass through their water vapor barrier. One can reasonably question whether creationists have ever considered the potential hazards of a continuous buildup of heat for thousands of years under their superheated steam blanket.

That, I propose, is a critical problem in the creationist model. They are so preoccupied with the capture and retention of solar energy and its presumed equal distribution everywhere,[7] that they see no need for escape valves. Nor do they recognize the possibility of dangerous imbalance and a lethal buildup of heat.

Creation folk scientists who reject the necessary balance of heat and other forces that sustain life have invented a lethal world. Fortunately, their heat machine was only sketched on paper and never had a trial run in the real world.

Notes

1. J. Ellis and T. H. Vonder Haar, "Zonal average radiation budget measurements from satellites for climate studies," *Atmospheric Science Paper* no. 240 (Fort Collins, Colo.: Colorado State University, 1976).

2. Ibid., p. 48.

3. The crossover at about 36°N in fig. 8 is not a permanent boundary over which energy flows from regions of energy excess. I estimate from zonally averaged data for the northern latitudes the following crossover points for months January to June, respectively: 15°, 22°, 36°, 49°, 63°, and 84°. In addition to the effects of season, the locations of land and water masses, direction and heat content of winds and ocean currents, and the deserts, steppes, mountains, forests, and other factors all determine the energy balance that satellites measure. Thus the E/A balance progresses annually from south to north and to the south again in keeping with available sunlight and the physical forces and features of the globe.

4. H. M. Morris, *Scientific Creationism*, General edition (El Cajon, Calif.: Master Books, 1974), p. 210n.

5. H. M. Morris, *The Beginning of the World* (Denver, Colo.: Accent Books, 1977), p. 77.

6. J. C. Whitcomb and H. M. Morris, *The Genesis Flood* (Phillipsburg, N.J.: Presbyterian and Reformed Publishing Co., 1961), p. 240.

7. Ibid.

16

Creationism's Fabricated Greenhouse

Aspects of Intuitive Meteorology

The minds of our sapient species often rest on first impressions and have too little time for critical thinking. We lay out small plots of ground and decide that the earth is flat. We declare the sun to be earth's orbiting satellite, for we have watched its daily transit.

We also watch the sky and reach certain conclusions about water. For instance, a heavy cloud moves across the sun and we perceive, correctly, that a high density of a certain form of water hinders light transmission.

Unfortunately, intuitive certainties may not always be true and if not critically examined may lead to serious error. The following analogy illustrates that hazard and raises a key issue.

An observer watches droplets of steam rise from a boiling pool and disappear as vapor into space. The vapor does not cast a shadow as does a cloud or a steaming geyser; it is invisible. The observer concludes that water in vapor form does not hinder the passage of light to earth.

That conclusion defines an elementary concept of creationist doctrine. For the past thirty years, it has been expounded vigorously by Henry Morris and John Whitcomb.

Why would these authors turn the water, allegedly uplifted at creation, into a water vapor greenhouse? The main reason, perhaps, was that they could not expect liquid water to remain in the sky and be available for the biblical flood. Moreover, they assumed that the vapor shell could be maintained unmixed above heavier gases and also that it would be transparent.

Invisible Vapor: Assumed Transparency

In speculating about water in gaseous form, Whitcomb and Morris apparently formed an early fixation on the word "invisible." Consequently, when they mentioned water vapor, it was nearly always the "invisible water vapor." For example, in discussing imagined preflood conditions they stated that, "The upper waters did not, however, obscure the light from the heavenly bodies and so must have been in the form of invisible water vapor."[1]

Finding that phrase again and again in their writings, one can only wonder why they keep reminding readers that people can't see water molecules. Was there a special meaning they intended to convey? Whitcomb's assumptions about conditions prevailing under the alleged preflood sky and 40-day rainfall, gave a partial answer, "This can be nothing less than the collapse of a stupendous transparent vapor canopy. . . ."[2] We infer from their statements that these authors took the words "invisible" and "transparent" to be synonymous.

This was confirmed by Henry Morris, who stated several assumptions about the canopy's nature and function. He told his readers,

> The "waters above the firmament" must have been in the form of invisible water vapor, extending far into space. They provided a marvelous "canopy" for the earth, shielding it from the deadly radiations . . . and producing a wonderful "greenhouse effect," sustaining a uniformly warm, pleasant climate all around the earth. Being invisible, these water vapors were of course transparent to the light of the heavenly bodies which were to be established on the fourth day.[3]

His "of course" reveals the basic assumption. For Morris, the matter was settled; water vapor, being invisible, must also be transparent.

The above quotations reveal serious conceptual and logical errors that may be expressed in a simple proposition: Invisible things are transparent; water vapor is invisible; therefore, water vapor is transparent.

To answer the false logic, we first admit that water vapor is invisible. But the major premise, invisible things are transparent, is false. Many microscopic entities such as bacteria, spores, viruses, organic and inorganic molecules, including water vapor, are invisible to human eyes, but they are not transparent. They are opaque to varying degrees, depending on their concentration and the wavelength of visible radiation.

Here we observe how creationists build on the misconception that invisibility is a valid criterion for judging transparency. Confusing the meanings of "invisible" and "transparent" has led to false conclusions that have become an integral part of their upside-down pyramid.

Gases and Greenhouses

A growing concern over the past half-century is the increasing level of infrared-absorbing substances that trap heat in the atmosphere and prevent its escape to space.[4] The principal factors responsible for introducing substances to the atmosphere are winds, ocean spray, lightning, fires, volcanic and tectonic releases, industrial processes and products, respiring organisms, and decaying matter. The substances introduced are dust, salts, ash, smoke, soot, and numerous gases, including volatile organic and inorganic molecules from respiration and decay.

Carbon dioxide has received much attention as a heat-trapping agent because of its steady increase above naturally produced levels, due to human-related activities. However, the effects of water vapor in the atmosphere are far more important than carbon dioxide in the heat exchange processes between atmosphere and earth. Creationists try to exploit this fact in developing their water vapor story. Whitcomb and Morris recognized that a relatively small quantity of water in today's atmosphere has an important role in "regulating earth's temperature."[5] But that small amount of water could never control solar heat and provide creationism's globally warm climate. Therefore, Whitcomb and Morris proposed their fundamental thesis.[6] A massive greenhouse blanket of water vapor would have greatly exceeded the paltry effects of today's atmospheric vapor, ". . . with a larger percentage of the sun's incoming radiant energy being absorbed and retained and uniformly distributed over the earth than at present, both seasonally and latitudinally." Latitude and season had little importance for these authors. Unlike the present levels of water vapor, the alleged preflood vapor greenhouse, in their view, would virtually have cancelled all effects of season and latitude on the global climate.

The Morris-Whitcomb greenhouse was simply a universal structure roofed over by a continuous dome of water vapor supported by heavier gases. The water vapor would absorb incoming heat. It would also prevent heat escape just as a greenhouse roofed with glass confines heat under the glass. The picture that emerges for creationists is one of universal warmth, humidity, light, and luxurious growth inside a "wonderful greenhouse."

Creationism's Greenhouse "Glass": Clear or Opaque?

Creationists need to face squarely the problems of their vaporous greenhouse structure. Could it have functioned as they have described it? More specifically, could the greenhouse glass (their alleged vapor dome) have been transparent? Or would it have been dangerously opaque and a threat to life on earth? To answer that question we should first try to determine how the transparency notion arises in people's minds and how it might be reinforced and perpetuated. My earlier analogy of a casual observer watching steam evaporate to space was intended to illustrate the folly of uncritically embracing first impressions.

Understandably, students with little knowledge of the chemical and physical properties of water, particularly its optical properties, might assume water vapor to be transparent. Persons not aware of the difference between invisible and transparent might also guess that invisible water molecules could not block sunlight. They might conclude, therefore, that a very great quantity of such molecules could envelop earth without reducing light transmission.

Apart from the lack of knowledge about water and light relations, certain stated opinions—even some practices by scientists—contribute to the impression that water molecules pose no barrier to illumination. One factor, undoubtedly, is the unqualified assertion that water vapor is transparent. I confess to a deep bewilderment upon hearing such a statement for the first time.

Scientists often use such words as "virtually," "almost totally," or other qualifying terms rather than deal in arbitrary absolutes.

Another factor that may strengthen the notion of complete transparency of water can be the dimensions of a graph on which a scientist plots light absorption data. Consider this analogy: a photographer's snapshot of 300-foot trees hardly shows 3-foot trees nearby, and the 3-inch seedlings in the area do not appear on the photograph. Obviously the camera was not focused on 3-inch plants. Likewise, scientists who focus on prominent water absorption bands (e.g., the infrared region of the spectrum) may fail to show any absorption in the visible region. Kondratyev called attention to the absence of known absorption bands for water vapor in one of his illustrations, "Figure 3.3 does not show any water absorption bands in the visible range, but they are present, though very weak, and are known as 'rain' bands within the wavelength interval of 572 to 703 mμ."[7] Kondratyev apparently was concerned that readers might be misled by his failure to show water vapor absorption on his published figure.

Other writers have presented detailed resolution of light absorption bands by vapor, but still did not resolve all absorption lines.[8] Houghton also commented on light absorption that doesn't completely disappear between the dominant lines and bands of absorbing gases:

> Molecular gases exhibit band spectra, each band consisting of a large number of closely spaced lines; several well separated bands are commonly found. Thus the spectra of gases appear to be discontinuous, although the absorption does not quite vanish between the lines or bands.[9]

The most complete record of absorption lines and bands that show water absorption of light is the tabulation by Moore, Minnaert, and Houtgast.[10] Over the spectral range of 541.4 to 748.0 nanometers, the atmospheric water vapor lines exceed 1,000. I counted more than 570 other lines that Moore et al. named "Atm" and described as "probably due to the water vapor molecule."[11] Sixteen water absorption bands are shown in the appendix table.

The human eye and mind cannot sense the difference that 10 percent and 80 percent atmospheric relative humidity has on the amount of light reaching the eye. Nevertheless, I assume that the notion of water vapor transparency will continue as intuitive certainty.

"Vapor Blanket" Water versus Current Estimates

Biblical writers and modern creationists have held tenaciously to the idea that the quantity of water in their preflood sky had to be immense. Rain fell night and day for forty days and created a flood that covered the highest mountains (Genesis 7:12, 19). Recent creationist authors assert that a collapsing vapor dome (plus exploding "fountains") stripped earth's vegetation, killed animals, and buried the debris—most of it in deep sediment. Creationist leaders have transformed the Genesis "waters above the firmament" into superheated steam that later condensed and fell to earth to become a large part of the ocean.

Noted earlier (chapter 2) was Lammerts's expectation ("most certainly") that the collapsed canopy would have diluted ocean salt concentration "perhaps by one-half."[12] To these authors the half-ocean volume of canopy water apparently seemed plausible enough to enhance the story they were developing. Another estimate of precipitable canopy water by these authors would have increased the volume of a presumed former ocean by 30 percent.[13] Dillow estimated for his precipitated canopy a volume of nearly 0.5 percent of the present ocean.[14]

The average total water content in the present atmosphere would equal about a one-inch depth if spread over the earth. How do creationist preflood estimates for canopy water compare with the present atmospheric water content? The answer is that vapor equal to 0.5 percent of the ocean would be 480 times the present atmospheric water. Vapor equivalent to 50 percent of the ocean would be 51,000 times the present atmospheric water.

Creationism's fictitious vapor shell is an absurdity existing only in the minds of its inventors and those who believe the story.

Notes

1. J. C. Whitcomb and H. M. Morris, *The Genesis Flood* (Phillipsburg, N.J.: Presbyterian and Reformed Publishing Co., 1961), p. 215.

2. J. C. Whitcomb, *The World That Perished* (Grand Rapids, Mich.: Baker Book House, 1973), p. 34.

3. H. M. Morris, *The Beginning of the World* (Denver, Colo.: Accent Books, 1977), p. 25.

4. S. Weisburd, "Greenhouse Gases En Masse Rival CO_2," *Science News* 127, no. 20 (1985): 308; R. M. White, "The Great Climate Debate," *Scientific American* 263, no. 2 (1990): 36.

5. Whitcomb and Morris, *The Genesis Flood*, p. 240.

6. Ibid.

7. K. Y. Kondratyev, *Radiation in the Atmosphere*, International Geophysics Series, vol. 12 (New York: Academic Press, 1969), p. 109.

8. M. D. King, D. M. Byrne, J. A. Reagan, and B. M. Herman, "Spectral Variation of Optical Depth at Tucson, Arizona, between August 1975 and December 1977," *Journal of Applied Meteorology* 19, no. 6 (1980): 726.

9. H. G. Houghton, *Physical Meteorology* (Cambridge, Mass.: MIT Press, 1985), p. 114.

10. C. E. Moore, M .J. G. Minnaert, and J. Houtgast, "The Solar Spectrum 2935Å to 8770Å," *National Bureau of Standards Monograph 61* (Washington, D.C.: U.S. Government Printing Office, 1966).

11. Ibid., p. xix.

12. Whitcomb and Morris, *The Genesis Flood*, p. 70.

13. Ibid., p. 326.

14. J. C. Dillow, *The Waters Above: Earth's Pre-Flood Vapor* (Chicago: Moody Press, 1982), pp. 136–37.

17

Creationist Preflood Era
Was There Light and Life?

Light Transmission and Attenuation

Several reactions between light and water are invariable. Increases of water molecules in the path of incoming light will progressively block more and more of the solar wavelengths necessary for life. The interference with free passage of light is called attenuation, a thinning out of light rays resulting from absorption and scattering.[1]

As the quantity of atmospheric water increases, the far-red wavelengths are absorbed more than the blue and other shortwave spectra. With further increases of water vapor in the atmosphere, or depth of liquid water, the quenching of light progresses steadily toward the blue and violet end of the spectrum. At great enough optical depth, even "blue" photon rays converge to extinction. The lifetime of a photon is the length of its path. Beyond the end of its path is darkness.

Water Vapor Absorption of Far-Red Light

Radiation research provides a rich legacy for science in the fields of light and water interactions and photobiology. Coulson compiled an excellent summary on "Milestones in Radiation Research," which lists discoveries that have occurred since the early 1600s.[2]

Early in this century, pioneering scientists in the Smithsonian Institution were developing a sophisticated technology for solar radiation measurements.[3] They also collected data on light absorption by water vapor.[4]

Among the highlights of research Coulson noted for 1908 was the "start of study of transmission of water vapor by F. E. Fowle."[5] In a remarkable endorsement, McDonald recorded that in the previous forty years of solar absorption studies

> . . . all but a single effort to compute free-air insolation [incoming solar radiation] absorption . . . depend ultimately upon the observational work of a single investigator, F. E. Fowle (1912, 1915).[6]

Fowle's 1915 data for vapor absorption of far-red light are plotted in figure 10.

If several times the 8 grams/cm^2 shown in figure 10 could remain in the sky, what water quantity in the vapor column would block all transmission of far-red energy? The regression line indicates that 51 grams of water vapor would have absorbed all the 700–740 nm light. In other words, water vapor equivalent to a half-meter liquid depth would block passage of virtually all far-red light through creationism's vaporous atmosphere. A half-meter of precipitable water vapor is only a small fraction of the water allegedly held in creationism's canopy model.

Interposing additional water between earth and the sun would not only block out all the 700-740 nm light, it would also reach far into the red wavelengths to limit their activity. Another way of testing the validity of creationist assumptions about light is to consider light-limiting effects on the physical environment and on living things.

Vital Relations: Light and Plant Pigments

The educated public generally understands that an interplay of hereditary and environmental factors determines the structural development and behavior of organisms. The genetic constitution of a typical plant sets limits on the organism's development and reactions. For example, the planted seed of a flowering land plant, after taking up water, follows controlled sequences that mobilize

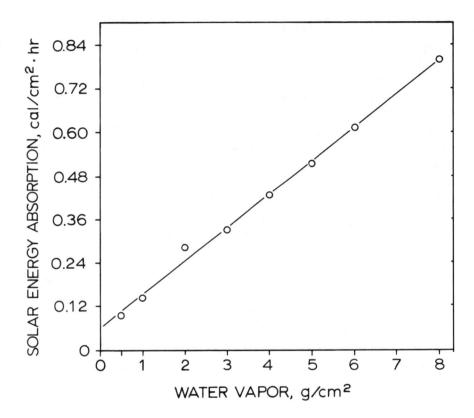

Fig. 10. Water vapor absorption of solar energy in the visible band, 700 to 740 nanometers. The estimated solar flux for the far-red band at the top of the atmosphere is 4.7 cal/cm² · hr. The depth of water vapor for each quantity shown (precipitable grams) can be visualized as a column one square centimeter in cross section extended to a region in space where absorption by water molecules was not detectable. Plotted values are those of J. E. McDonald, "Direct Absorption of Solar Radiation by Atmospheric Water Vapor," *Journal of Meteorology* 17 (1960): 321. Values were computed from Fowle's transmission curve: F. E. Fowle, "The Transparency of Aqueous Vapor," *Astrophysical Journal* 42 (1915): 409, fig. 4.

seed nutrients into embryonic root and leaf tissues. Stems elongate, and chlorophyll appears in leaves exposed to light. Days or perhaps months later, the plant may bloom, set fruit, and produce seed.

The environment to which a plant is subjected also sets limits upon the plant's development. Important external factors that influence plant growth include the following: water in soil and air, temperature of soil and air, nutrient solutes in soil, gases in air and soil, and radiant energy in the needed quantities and wavelengths.[7]

Earth's most celebrated photochemical activity is the combining of carbon dioxide and water in plants through reactions between sunlight and chlorophyll. Chlorophyll captures sunlight mostly in the wavelengths of blue-violet and red and might be called earth's workhorse in food production. Nobel estimated that yearly "about 7.2×10^{13} kg (72 billion tons) of carbon are fixed into organic compounds by photosynthetic organisms."[8]

Another pigment, phytochrome, has a very different but vital role. It is a sensor that triggers many plant responses. Among the responses attributed to phytochrome action are seed respiration, seed germination, fern spore germination, internode elongation, plumular hook unfolding, leaf enlargement, flower initiation, flower development, sex expression, dormancy of buds, and leaf abscission.[9]

After Fowle's early measurements of far-red light, it was several decades before scientists irradiated plants with wavelengths in the 600–800 nanometer range. They discovered that both far-red (R_f) and red (R) radiations were involved in the action of phytochrome. However, exposing plants to varied ratios of R_f/R caused very different responses and growth patterns. Review articles by Hendricks and Borthwick[10] and Toole[11] give data on phytochrome action and on factors that affect its activity in plants.

Without phytochrome activity initiated by far-red light, plants could not respire. Plant seeds could not use their food reserves for seedling growth. There could be no germination, stem growth, leaves, chlorophyll, flowers, or seeds. To be active, phytochrome in a plant organ (seed, stem node, leaf) must have been irradiated within effective time intervals by appropriate energy quantities of R_f and R light quanta.

This raises questions for creationists to answer. How did phytochrome-containing plants exist in their preflood world if a massive vapor canopy prevented all R_f transmission to earth? Not only must creationists find scientific answers about far-red deprivation, they must face other potential deficiencies. If any plants could survive without far-red irradiation, how could they have grown and reproduced under a canopy that prevented the shorter, red to violet wavelengths from reaching earth?

Vapor Scattering of Light in the Red to Violet Range

From Fowle's experiments we have seen that atmospheric water vapor (in amounts much smaller than creationists have imagined in their vapor canopy) can block far-red transmission. The Smithsonian Institution investigated other wavelengths. Over a period of several years, Fowle measured a broad range of spectra (342 to 2,348 nm) transmitted through cloudless atmosphere above Mt. Wilson in California. As noted by List, Fowle's published values represent "only the effect of scattering by 1 centimeter of precipitable water vapor" in the vertical path.[12] In table 2 below are data for six visible wavelengths selected from Fowle's table 1.

Fowle's results are clear and undeniable; light entering and light emerging from a space that contains water vapor cannot be equal. Even 1 gram of water vapor causes a measurable decrease in light transmission.

Henry Morris, Ph.D., hydrologist, engineer, author, professor, and formerly chairman of the civil engineering department at the Virginia Polytechnic Institute and State University, is now president of the Institute for Creation Research. It is difficult to understand how such training and experience could have deprived him of appreciation for basic interactions between light and water vapor in the atmosphere. Whether from neglect of the classic literature on physical and optical properties of water or a refusal to consider such information important, I cannot say. In any case, creationists Morris and Whitcomb are totally incorrect in their notion about vapor transparency and in their conclusion that virtually unlimited atmospheric vapor could have been held in the sky without hindering light transmission.[13]

Table 2. Light attenuation from scattering by water vapor. Transmission values, selected over the range of violet to red wavelengths, are from Fowle (F. E. Fowle, "The Transparency of Aqueous Vapor," *Astrophysical Journal* 42 [1915]: 403). Precipitable vapor in the vertical air column was 1 g/cm². The inserted line of values for all visible wavelengths (*) represents creationism's unverified assumption of total water-vapor transparency.

Wavelength (nm)	Transmission (T)	Attenuation (Scattering) (a_{wv})	Opacity (1/T)	Optical Density (OD)
384	0.945	0.055	1.058	0.0246
413	.953	.047	1.049	.0209
452	.961	.039	1.041	.0173
503	.968	.032	1.033	.0141
574	.970	.030	1.031	.0132
686	0.981	0.019	1.019	0.0083
*—	1.000	0.000	1.000	0.0000

Key: Transmission (T) = ratio of transmitted to incident light (L_t/L_i), where L_i = 1. Attenuation by water vapor (a_{wv}) = 1 - T. Opacity is the reciprocal of T = 1/T = L_i/L_t. Optical Density (OD) = log (L_i/L_t).

Light Penetration Compared: Vapor versus Liquid Water

All bodies of clear water absorb and scatter solar energy, so that near the 100-meter depth only the most penetrating wavelengths persist, and at very low energy. For example, Jerlov reported that at 100 meters the total solar irradiance narrows to a spectrum of blue light with energy of about 0.1 watt per square meter.[14] Jerlov also gave percentages of downward irradiance as far as 200 meters in clear ocean water.[15] The energy source was direct solar and "sky" radiation over the 300–2,500 nanometer spectrum. In his paired values given below, the first is the meter depth; the second is percent of the surface irradiance: 0, 100; 1, 44.5; 10, 22.2; 50, 5.3; 100, 0.53: 150, 0.056; 200, 0.0062.

Only 44.5 percent of the incident radiation penetrated beyond 1 meter. The loss of the 55 percent in the surface meter can be accounted for by virtual extinction of all infrared, as well as depletion of the far-red and considerable loss of red wavelengths shorter than 700 nanometers. Increasing the water depth to 10 meters and deeper would have progressively blocked shorter wavelengths and further isolated a narrow spectrum of blue light until, at great depth, all light was extinguished.

Whitcomb and Morris no doubt appreciated the above facts when writing their water vapor story. They understood, of course, that liquid water could not be held in the sky; they knew that liquid water, in the great quantity they imagined, could not transmit light for biological needs. These authors also fixed their minds on the notion that water vapor molecules were invisible and guessed that very great concentrations of vapor in the sky would be transparent to incoming light. Perhaps to these writers it seemed reasonable, therefore, to reject liquid water and select its vapor as the biblical medium in their preflood sky.

Could water vapor equivalent in mass to a layer of liquid water have scattered less light than the liquid medium? Several references on the basic physics of light attenuation show that water vapor is much less transparent than liquid water.[16] The classical reference on physical optics and the relative scattering of light by liquid and vapor is by Wood:

> Now a very striking thing is shown in the case of scattering by a liquid and its vapor. We might very reasonably expect the scattering of the light to be proportional to the number of molecules present, and as there are roughly 1000 times as many molecules in a given amount of liquid as in an equal volume of its vapor at atmospheric pressure, we might look for an intensity [of scattered light] 1000 times as great in the liquid as in the vapor. As a matter of fact we observe an intensity less than 50 times that of the vapor.
>
> This is, however, precisely in accord with the theory, for in the liquid we have a nearer approach to a continuous medium and if the medium were perfectly continuous and structureless there could be no scattering. . . .[17]

Water vapor molecules in the atmosphere move randomly and freely. Further discussion on the physics of light scattering can be found in *The Feynman Lectures on Physics*.[18]

One might reasonably ask: If Morris and Whitcomb had known that vapor is less transparent than liquid water, would they have taken so much time expounding on their liquid-to-vapor transformation and vapor transparency?

Creationists fail to bolster assumptions with scientific evidence. They build their assumption pyramid upside down. They begin with a lower-tier stone called "ocean in the heavens," upon which they place another called "transparent vapor." Both stones are products of surmise and are supported only by specious rhetoric. They solve no problems for creationists. The vapor "transparency" notion provides no rational base for arguments about their illuminated vaporous greenhouse and the assumed abundant life within it.

Notes

1. Quantities of light absorbed and scattered depend on several factors: the spectral wavelengths, the electronic properties of molecules, the size of molecules, and the density of molecules or other matter in volumes of space. Water vapor absorbs relatively more far-red and red than it does blue spectra. A great vapor density would absorb all light. Shortwave spectra (blue-green and blue-violet) scattered earthward account for the visible blue of the sky. Shortwave spectra scattered back to space are part of earth's albedo. See fig. 7 and legend, path D, for estimated annual energy quantities.

2. K. L. Coulson, *Solar and Terrestrial Radiation* (New York: Academic Press, 1975), pp. 11–21.

3. F. E. Fowle, "The Spectrometric Determination of Aqueous Vapor," *Astrophysical Journal* 35 (1912): 149–62; F. E. Fowle, "The Determination of Aqueous Vapor Above Mount Wilson," *Astrophysical Journal* 35 (1913): 359–72.

4. F. E. Fowle, "The Transparency of Aqueous Vapor," *Astrophysical Journal* 42 (1915): 394–411.

5. Coulson, *Solar and Terrestrial Radiation,* p. 18.

6. J. E. McDonald, "Direct Absorption of Solar Radiation by Atmospheric Water Vapor," *Journal of Meteorology* 17 (1960): 319.

7. Another factor is *geotropism,* a display in which gravity serves to orient root and shoot growth in characteristic up or down direction. Earth's orientation to the sun sets temporal limits on radiation exposure, and causes regional heating, pressure differences, and wind activity. Other important factors are soil and air pollutants, diseases, competition, and parasitism.

8. P. S. Nobel, *Biophysical Plant Physiology and Ecology* (New York: W. W. Freeman, 1983), p. 239.

9. S. B. Hendricks and H. A. Borthwick, "Control of Plant Growth by Light," *Environmental Control of Plant Growth* (New York: Academic Press, 1963), p. 242.

10. Ibid., pp. 233–63.

11. V. K. Toole, "Effects of Light, Temperature and Their Interactions on the Germination of Seeds," *Seed Science and Technology* 1 (1973): 339–96.

12. R. J. List, *Smithsonian Meteorological Tables,* 6th rev. ed., vol. 114 (Washington, D.C.: Smithsonian Institution, 1963), p. 432.

13. J. C. Whitcomb and H. M. Morris, *The Genesis Flood* (Phillipsburg, N.J.: Presbyterian and Reformed Publishing Co., 1961), pp. 215, 256; H. M. Morris, *The Beginning of the World* (Denver, Colo.: Accent Books, 1977), p. 25.

14. N. C. Jerlov, *Marine Optics,* Elsevier Oceanography Series, no. 14 (Amsterdam: Elsevier Scientific Publishing Co., 1976), figure 75.

15. Ibid., table 28.

16. Personal communication (October 11, 1990) from Robert S. Fraser, Goddard Space Flight Center, NASA, Greenbelt, Maryland. He affirmed the greater transparency of liquid water relative its vapor and cited authorities on the subject. See references in the text.

17. R. S. Wood, *Physical Optics* (New York: Dover Publications, 1967), p. 428.

18. R. P. Feynman, R. B. Leighton, and M. Sands, *The Feynman Lectures on Physics,* vol. 1 (Reading, Mass.: Addison-Wesley, 1963), pp. 32-6 to 32-9.

18

Creationism's World
Perpetually Bright or Dark?

From sheer imagination creationists generate impressive lore and construct inverted pyramids of fiction. Consider the following imaginary world embellished totally with fanciful rhetoric:

> The lands that once had teemed with animals and people and lush vegetation now were barren and forbidding. The air which formerly was warm and still now moved in stiff and sometimes violent winds, and there was a chill on the mountain slope where the Ark rested. Dark clouds rolling about the sky, which had once been perpetually and pleasantly bright, seemed to threaten more rains and another flood.[1]

Thus, a creationist visionary abandons the real world and sees it as he wishes it to be. There never was a time in earth's history when the sky was perpetually bright. The orbiting and revolving earth makes that impossible.

If we believe literally what some creationists have published, their preflood sky could never have been bright, but would have existed in perpetual darkness. In the following pages I show how water vapor pressure and densities would dictate that darkness.

Vapor Pressure and Light Absorption

Most people can appreciate—at least in elementary ways—what is meant by compression of air and the resulting pressure in some confined volume. A four-year-old blows up a toy balloon. A twelve-year-old inflates a bicycle tire using a simple hand pump or other piston-and-cylinder apparatus. External force on the piston drives gas molecules into a smaller volume where they exert pressure on all cylinder walls and the piston surface.

In contrast, earth's atmosphere is open to space. It does not have a driving piston but is subject to gravity. Thus, at sea level the weight of its gases registers globally a pressure that does not vary widely.[2]

A column of air does not have rigid walls; instead, "walls" of ambient gases surround it on all sides. In such an open volume, each gas exerts in all directions a pressure that is not dependent upon which gases are present.

The amounts of a few gases can vary considerably in different regions of the globe. For example, water vapor may be low over a desert but high over a rain forest. Let us imagine a vertical column of water vapor in a cloudless sky. The column has a cross-section of one square centimeter and extends to a point in space where water molecules are not detectable.

Kondratyev stated that in nature the vapor pressures in such columns usually range from tenths of a gram to 1 or 2 grams per square centimeter and that solar radiation absorbed is on the order of tenths of a calorie per square centimeter per minute.[3] He emphasized, however, that the "total extinction of radiation in the atmosphere is far greater" because of scattering of light energy. In his table 3.8, Kondratyev shows the effects of vapor pressure on light absorption.[4]

I emphasize the following five points: (1) the density of water vapor increases markedly with increased pressure, (2) light transmission decreases with increased water vapor density, (3) very high vapor density blocks all transmission of light, (4) total blockage does not occur in the real world, because too little vapor is in the sky, and (5) blockage could have occurred under regimes of some vapor models the creationists have imagined.

Sources for vapor temperature/pressure values are widely

available.[5] Another important source[6] updated earlier steam tables and gave vapor densities.[7]

Comparisons: Vapor Canopy Models

In comparisons below, I present quantities of alleged water vapor in three creationist models. I then show the volume increases in the ocean volumes and depths resulting from the alleged precipitation of the canopies in the Genesis flood. The three scenarios and related Steam Table data (table 3) will be discussed as Canopy One, Canopy Two, and Canopy Three.

Canopy One. A vapor canopy proposed by Dillow contained 40 feet of precipitable water.[8] At the canopy's lower boundary (30,000 feet, suggested by Dillow[9]) the vapor temperature would be 104.4° C (220° F). At that boundary the vapor pressure would equal about 89 centimeters of mercury or a force of about a 1.2 kilograms per square centimeter. Dillow[10] extracted temperature and pressure values from the 1967 *Steam Tables.*[11] The canopy allegedly would have held nearly 500 times the total water in today's atmosphere. And when it collapsed it would have increased the ocean 1.5 million cubic miles.

A reliable statement about temperature, pressure, or density values at some lower boundary of a global vapor canopy would require evidence that such a boundary could exist and be maintained at that level above earth. Creationists have assumed that vapor stability was possible for at least 1,656 years. Incredible control over matter and natural processes would have been necessary to achieve that stability. In their minds, creationists lowered high mountains, abolished turbulent winds, eliminated injections of dust from volcanoes or from outer space, kept the vapor super-heated, and denied Dalton's law, which states that gases diffuse from regions of high concentration to regions of lower concentration.

Lucid discussions of pressure and density are possible only when boundary conditions can be determined. In contrast, one has to step into successive mental traps to discuss pressure or density at a surmised boundary of an imagined vapor canopy, allegedly held above earth throughout a mythical preflood time. Moreover, the notion that any creationist vapor blanket could be preserved until a preordained time for its collapse (Genesis:

Table 3. Three creationist vapor canopy and ocean water models with related Steam Table data. VC/TWP is the projected ratio of vapor canopy water to the total water presently in the atmosphere. Steam data are from L. Haar, J. S. Gallagher, and G. S. Kell, *NBS/NRC Steam Tables: Thermodynamic and Transport Properties and Computer Programs for Vapor and Liquid States of Water in SI Units* (New York: Hemisphere Publishing Corporation, 1984), pp. 5–15.

	Creationist Water Models			Steam Table Data			
	Ocean Increase		Aerial Water		Water Vapor		
No.	Volume	Depth	Ratio	Temp	Pressure		Density
	$10^6 mi^3$	ft	VC/TWP	°C	cm Hg	kg/cm^2	g/cm^3
1	1.5	40	480	104.4	89.7	1.2	0.0007
2	73	1,950	23,400	274	4,390	60	0.030
3	158	4,250	51,000	328	9,404	128	0.074

6:17, 7:4) is a subject for religious surmise and faith, not a topic for scientific testing and confirmation.

Canopy Two. This model was proposed by Whitcomb and Morris, who assumed ". . . a sudden increase of 30 per cent in the volume of the ocean. . . ."[12] The 1,950-foot increase in ocean depth could not have covered their highest mountains. Nor could light have passed through vapor equivalent to a 1,950-foot depth of water.

Canopy Three. This scenario also was supported by Whitcomb and Morris.[13] They borrowed the half-ocean estimate (of canopy water that diluted ocean saltwater) from Walter Lammerts, a geneticist and assistant professor of horticulture at the University of California at Los Angeles. The liquid depth of 4,250 feet and the 158 million cubic miles are the most fantastic estimates yet offered for precipitated vapor. Those quantities, however, would still not have been adequate to cover 7,000-foot mountains without the additional releases of alleged cavern sources and uplift of ocean floors proposed by these authors. The half-ocean volume of collapsed vapor, nevertheless, must have seemed adequate for their geological assumptions, which included worldwide transport of eroded materials and deep burial of organic matter.

These authors had other roles in mind for the massive vapor blanket. They imagined that it would greatly outperform the paltry amount of vapor in the present sky by capturing solar heat and uniformly warming the globe. They also assumed that light would penetrate their vapor blanket. Unfortunately, the quantity of vapor would have been far too great to allow passage of light.

If Morris and Whitcomb believed that the sun, moon, and stars were created to give light on earth, why did they invent conditions that would have prevented light from reaching earth? Apparently these authors constructed their vapor greenhouse without knowing that it could never be illuminated.

Crucial Words: Water, Mountains, and Light

Most fundamentalist Christians defend the Bible as the infallible "Word of God." For them the biblical scriptures are trustworthy in every word and in every matter with which they deal.

In the introduction to their book, Whitcomb and Morris stated their desire "to ascertain exactly what the Scriptures say about the Flood and related topics."[14] As for written words, they embraced the "doctrine of verbal inerrancy of Scripture" and quoted Warfield about supernatural control extending even "to the choice of words by the human authors (verbal inspiration). . . ." Another stated conviction was that "complete divine inspiration" imparts "perspicuity of Scripture,"[15] meaning that ideas expressed were clear and understandable.

How closely do creationist writers hold to their stated precepts and convictions? Below, I consider some rejections and contradictions of explicit Bible words.

Contradictions About Water. One of the first statements about water in the Bible tells of liquid water being separated and lifted into space (Genesis 1:6, 7). Dillow argued repeatedly[16] that the plain language does not refer to clouds or mist or fog, but to a liquid ocean like that of ancient creation myths of the Near East.[17] He maintained that "Genesis teaches us that it [the biblical water canopy] was a liquid ocean phase and not cloud, ice, or water vapor."[18] Since the identification of ordinary, visible water was explicit and understandable, one could assume that the simple term "water" was an example of creationism's "verbal inerrancy" and should, therefore, be accepted as undeniable truth. Should it be accepted as truth?

Dillow considered it "scientifically impossible to account for the canopy if it remained in the liquid form."[19] It was possible to hold it aloft, he believes, only if it existed as vapor. He thus offered his vapor surmise, "a theory . . . based on the significance of the words of Moses for modern science—but a theory not explicitly taught in Genesis."[20]

I presented Dillow's conjectures about vapor in table 3 and in discussion of Canopy One.

Morris and Whitcomb similarly rejected the biblical word for liquid water—but without Dillow's bold contrast between Near East myths and modern concepts about vapor. These authors were burdened with concerns about massive amounts of vapor in the sky. Their Canopy Two held fifty times the water Dillow allowed in his model, and their Canopy Three held more than one-hundred times as much. They labored under the misconception that invisible water molecules were also transparent. Thus they fabricated a story in which neither starlight nor any other light could have reached earth.

Misconceptions: Water and Mountain Heights. Mountains that Morris and Whitcomb lowered were still too high to be covered by available water without raising ocean basins. Dillow's preflood model required continents that were virtually flat.

Dillow believed that the stability of his canopy above earth's atmosphere was due to an "apparent lack of major mountain ranges."[21] He further believed that Bible statements (Genesis 7:11, Psalm 104:8) indicated "major mountain building . . . during the Flood and immediately after it." He then recorded this amazing commitment:

> In the model to follow it will be assumed that during the pre-Flood era, there were *no* mountains. Certainly there may have been rolling hills, but no major mountains. Therefore the lower edge of the canopy (about 30,000 feet) would never be in danger of intersecting the landscape. Furthermore, the convective updrafts produced by wind's blowing against the sides of steep mountains would have been severely limited.[22] (Emphasis in original)

Dillow believed that eliminating atmospheric turbulence and diffusion was the only way to prevent rapid mixing with the lower atmosphere and destruction of the canopy. His strategy seems much like that of Whitcomb and Morris. If high mountains hinder the development of the story, remove them, or at least lower them to manageable heights.

Dillow gave 2346 B.C. as the year in which the Genesis flood and the first mountain building occurred.[23] In his view, therefore, the present mountains would be less than 4,340 years old. He also has stated that Bishop Ussher's 4004 B.C. estimate for earth's creation "was not far off."[24]

Misconceptions about Light. Dillow should be given credit for understanding that water vapor absorbs and scatters light. It is not transparent. "Furthermore," he emphasized, "more light will be blocked out by water in vapor form than water in liquid form"[25] (emphasis in original).

Genesis (1:16–17) is an ancient myth that the sun, moon, and stars were created and positioned in the sky to give light on a preexisting earth. One of Dillow's major concerns was that star-light must pass through his vapor canopy, for if stars were created to give light, then the ancients should have been able to see them. He therefore loaded page after page[26] with information on star magnitudes, intensities, attenuation, extinction, and where in the sky the stars could have been observed. Thus, Dillow could state, concerning his estimated 743 stars "potentially visible to Adam," that, "When Adam looked into the antediluvian heavens, assuming 40 feet of precipitable water in the vapor canopy, he saw about 255 stars on a clear night when the moon was dark. . . ."[27] However, he concluded that under a full moon, with fifth-magnitude stars obscured, Adam "would have seen only 209 stars."

I propose that purported visibility of stars under Dillow's vapor canopy is less important than whether plants could have existed under his canopy. He gave only brief recognition to the need for light by referring to plants flourishing under illumination of the "tungsten bulb, which is rich in red light."[28] Dillow's 40-foot canopy of water would have blocked out far-red and red light needed for the action of phytochrome and chlorophyll pigments in plants.

Consideration of plants and their specific needs is crucially important. Without plants we eliminate stargazers.

Fabrications that Backfire. Creationists face perilous challenges at the very base of their inverted pyramid. They start erecting the first tier of stones on what purportedly is the "Word of God." Actually, the foundation has obvious flaws that are beyond repair. Evidence for such flaws comes from creationist builders themselves; they must fashion rhetorical props to give their own work imagined stability.

One of the crucial dilemmas is how much credibility will creationists give to the celestial ocean myth. They must decide if they accept explicit words and the intended meaning of the biblical author who wrote about *liquid* water elevated as a celestial ocean (Genesis 1:6–7). The narrative reported that 1,656 years after creation (biblical time) the ocean spilled through heavens' floodgates and over the whole world. The author explained that "windows of the heavens were opened" and after forty days they "were closed" (Genesis 7:11, 8:1).

Competent scholars have concluded that the Genesis narratives are not scientifically reliable. Henry Morris and the ICR staff have the following to say about such conclusions:

> This type of Biblical exegesis [not treating Genesis as factual] is out of the question for any real believer in the Bible. It is the method of so-called "neoorthodoxy," though it is neither new nor orthodox. It cuts out the foundations of the entire Biblical system when it expunges Genesis 1–11. The events of these chapters are recorded in simple narrative form, as though the writer or writers fully intended to record a series of straightforward historical facts; there is certainly no internal or exegetical reason for taking them in any other way.[29]

Despite their professed devotion to sacred scriptures, some creationists apparently see themselves as authentic creators of a new canopy model that supersedes the biblical model.[30] Dillow stated that his "canopy theory" was not the teaching of Genesis or the Bible.[31] Could Whitcomb and Morris make the same statement? They admitted that they have not scientifically verified that their vapor canopy existed.[32]

Challenging ancient sacred myths can be risky. Modern vapor canopy advocates, the products of a civilization that offers advanced education, have produced stories with no more credibility than the water myth produced by the biblical authors. Creationists should not be criticized for declining to accept as fact—or even if they repudiate—the purported celestial, liquid ocean. Neither should they be praised for trying to improve the story with added fictions. They failed because myths can be amplified and adorned, but they never become reality.

Notes

1. H. M. Morris, *The Beginning of the World* (Denver, Colo.: Accent Books, 1977), p. 114.

2. Atmospheric pressure equals the sum of the partial pressures of constituent gases in the air. That pressure and the pressure of each gas can be expressed in terms of millibars (mb), millimeters of mercury (mm Hg), pounds per square inch (psi), grams per square centimeter (g/cm^2), atmospheres (atm), and pascals (Pa). The following equation allows conversion among pressure terms: bar = 1,000 millibars = 750.062 mm Hg = 14.5038 psi = 1,019.716 g/cm^2 = 0.986923 atm = 100,000 Pa.

3. K. Y. Kondratyev, *Radiation in the Atmosphere,* International Geophysics Series, vol. 12 (New York: Academic Press, 1969), p. 115.

4. Ibid.

5. R. C. Weast (ed.), *CRC Handbook of Chemistry and Physics,* 65th ed. (Boca Raton, Fla.: CRC Press, 1984), pp. E–16 to E–21, D–192 to D–194.

6. L. Haar, J. S. Gallagher, and G. S. Kell, *NBS/NRC Steam Tables: Thermodynamic and Transport Properties and Computer Programs for Vapor and Liquid States of Water in SI Units* (New York: Hemisphere Publication Corporation, 1984), pp. 5–15.

7. In defining the critical point for water, a widely used handbook (R. C. Weast [ed.], *CRC Handbook of Chemistry and Physics,* 65th ed. [Boca Raton, Fla.: CRC Press, 1984], p. F–74) gives v, the specific volume, incorrectly as a density value, 3.10 g/cm^3. The specific volume should be 3.10 cm^3/g vapor at the critical point. The correct vapor density (the reciprocal of specific volume) is 1 $g/3.10$ cm^3 or 0.322 g/cm^3.

8. J. C. Dillow, *The Waters Above: Earth's Pre-Flood Vapor Canopy* (Chicago: Moody Press, 1982), p. 248.

9. Ibid.

10. Ibid., p. 240.

11. The *Steam Tables* have been reproduced since 1967 in numerous editions of the *CRC Handbook of Chemistry and Physics* (Boca Raton, Fla.: CRC Press).

12. J. C. Whitcomb and H. M. Morris, *The Genesis Flood* (Phillipsburg, N.J.: Presbyterian and Reformed Publishing Co., 1961), pp. 124, 326.

13. Ibid., p. 70.

14. Ibid., p. xx.

15. Ibid.

16. Dillow, *The Waters Above,* pp. 49–51, 57–58, 63, 109, 221.

17. Ibid., p. 51.
18. Ibid., p. 111.
19. Ibid., p. 221.
20. Ibid., p. 222.
21. Ibid., p. 248.
22. Ibid.
23. Ibid., p. 163.
24. Ibid., p. 162.
25. Ibid., pp. 287, 294.
26. Ibid., pp. 287–304.
27. Ibid., p. 203.
28. Ibid., pp. 180–81.
29. H. M. Morris (ed.), *Scientific Creationism*, General edition (El Cajon, Calif.: Master Books, 1974), p. 244.
30. History reveals widespread editing of Bible stories by avowed believers. Authority for the "facts" then shifts from the ancient author to modern editor. A classic example is that of leading creationists who set aside the biblical food plan for survival of animals on Noah's ark. They substituted their own hibernation plan, but gave no evidence that a year's survival without feeding care was possible. Their "improved" myth was no more credible than the original.
31. Dillow, *The Waters Above*, p. 222.
32. Whitcomb and Morris, *The Genesis Flood*, p. 241.

Part Three

Public Concerns
and Responsibility

19

Religion and Politics

Religious Fundamentalism: Activity and Pervasiveness

The scientific community accepts religious tolerance and civility as praiseworthy qualities in society. Although scientists may reject creationist dogma and agenda, they are not protesting for rights to enter fundamentalist schools to deliver opposing views. They are not suing such institutions for equal time in classrooms. Nor do they demand adoption of curriculum guides or certification of their textual materials to achieve balanced treatment. Neither are mainline church members who hold liberal views of theology and social goals determined to force their views on others. The reason for such respectful behavior is simple. Americans devoted to constitutional principles are unwilling to deny freedoms guaranteed to every individual. They hold religious institutions of law-abiding citizens inviolable to trespass.

Unfortunately, inveterate religious zealots do not return respect for respect. They accept no physical bounds. They see all institutions as well as human bodies and minds as property on which they may freely trespass.

Fundamentalist Americans enjoy freedom to assemble, worship, teach, and practice numerous other religious rites. Their children receive instruction in church and Sunday schools, vacation Bible schools, and church camps. They establish missions, seminaries, institutes, and colleges. They conduct campus ministries through counseling, clubs, workshops, and crusades.

Moreover, their churches and related institutions produce vast quantities of religious artifacts and souvenirs. Their literature includes tracts, pamphlets, lesson aids, devotional guides, bulletins, magazines, music, poetry, and plays. Weightier productions are books of sermons, commentaries, theological treatises, lexicons, concordances, testaments, and Bibles in many versions and languages. Indeed, the religious materials they publish add to untold millions per year. The combination of religious assaults by radio and television plus that of the print media increases greatly the potential for continuous blanketing of society with religious sectarian opinion.

But despite guaranteed freedoms and such vast opportunity, militant fundamentalist leaders are not satisfied. They devise new political strategies and push for greater influence and power. The concept of a "Christian nation," where religious devotees control public institutions, is widely emphasized in Far-Right circles. Jerry Falwell asserted in a sermon on July 4, 1976, "The idea that religion and politics don't mix was invented by the Devil to keep Christians from running their own country."

Beverly LaHaye, leader of a powerful network of more than 560,000 women in Concerned Women for America, also calls for a heavy mix of religion in politics. She stated emphatically, "Yes, religion and politics do mix. America is a nation built on Biblical principles. Christian values should dominate our government."[1]

Undoubtedly, the greatest challenge for fundamentalist Christian leaders and their followers is the control of public education. An extreme statement of that hoped-for dominance has been expressed by Jerry Falwell:

> I hope I live to see the day when, as in the early days of our country, we won't have any public schools. The churches will have taken them over again and Christians will be running them. What a happy day that will be![2]

Robert Simonds has led in the nationwide establishment of more than 500 chapters of Citizens for Excellence in Education (CEE). He also imagines a time of complete Christian dominance in all of America's 15,700 school districts:

When we get an active Christian parents' committee (CEE) in operation in all districts, we can take complete control of all local school boards. This will allow us to determine all local policy; select good textbooks, good curriculum programs; superintendents and principals. Our time is come![3]

Paul Weyrich, president of Free Congress Research and Education Foundation and the leader in efforts to elect right-wing candidates to office, described in detail the Far-Right agenda:

When we take control of our states and localities, and we can point to mayors and sheriffs and county board members and school board members and state legislators and finally Congressmen, then, we can talk about advancing our agenda in this country and not until then.[4]

Political Apostles and Religious Reforms

Although the Reagan and Bush administrations didn't explicitly agree with all of the Far-Right's goals, common interests and notable cooperation in religious efforts were clearly evident.

Passing the Religious Test. The subject of religion occurs twice in the U.S. Constitution. It appears first in Article VI, which declares:

The Senators and Representatives . . . and all executive and judicial officers both in the United States and in the several States shall be bound by oath or affirmation to support this Constitution; but no religious test shall ever be required as a qualification to any office or public trust under the United States.

What obligations do ministers, priests, and other religious groups have toward political office seekers? Would it not be honorable and prudent policy to avoid religious tests "as a qualification to any office or public trust under the United States"?

Unfortunately, political candidates and campaign committees do not always abide by the constitutional standard. The 1984 Republican National Convention provides an example. Before the convention, Christian ministers and priests across Texas received a letter from Senator Paul Laxalt, chairman of the Reagan-Bush campaign. In quotes from that letter, a Lutheran minister and a Rabbi have presented an astonishing exposé of religious influence. Their comments and quoted parts of Laxalt's letter are as follows:

> "Dear Christian Leader.
> "As leaders under God's authority, we cannot afford to resign ourselves to idle neutrality in an election that will confirm or silence the President who has worked so diligently on your behalf and on behalf of all Americans," the letter said. It spoke of President Reagan's "unwavering commitment to the traditional values which I know you share," and ended with a call on the clergymen receiving the letter to "organize a voter registration drive in your church . . . to help assure that those in your ministry will have a voice in the upcoming elections . . . a voice that will surely help secure the re-election of President Reagan and Vice President Bush."
>
> Accompanying the letter was a flyer with the legend, "Christian Voter Program Information Enclosed," and a photograph of a smiling Ronald Reagan and the headline, "President Reagan Has Been Faithful in His Support of Issues of Concern to Christian Citizens."[5]

The testimonial letter certified that the president had been faithful and had worked energetically for concerns of the Christians. Obviously, he passed the Christian voter religious test and was ostensibly qualified for reelection to America's highest political office. Such deliberate strategy might be dismissed as a passing quirk in the political process, except that the courtship has continued into the present decade.

Is there indeed a "Christian voter religious test" that violates the spirit of the Constitution and renders ineffective its wording and intent? Would an incumbent president or other official who obligates himself or herself to establish religion (e.g., prayer in public schools), which the First Amendment prohibits, be in violation of the oath to support the Constitution?[6]

Politics and Schoolroom Religion

President Ronald Reagan's strong advocacy for prayer in all American schools was related apparently to what he imagined his own experience to be as a child growing up in Illinois.[7] He didn't "find any problem going to school and listening to different kinds of prayer." A problem, however, with Reagan's alleged "listening" is that spoken prayer was not allowed in his school. The report stated that the Illinois American Civil Liberties Union contacted Reagan's elementary school principal, Esther Barton, now retired in Dixon, Illinois, who said, "In my memory we never allowed prayer in my school or any other school in the area." A decision by the Illinois Supreme Court in 1910 had ruled that Bible reading and prayer in public schools violated the state constitution.

In three school-related cases in 1985, the U.S. Supreme Court ruled consistently against mixing religion and public education.[8] But in that year, as in preceding years, the Reagan-Bush administration was taking counterpositions to accommodate religion. Ronald Reagan, elected in 1980, had repeatedly expressed the assumption that God had been expelled from the classroom. In campaigning he had voiced support for official school prayer and advocated Bible reading and teaching. He disparaged evolution as only a theory. Most noticeable, the president made promises and appointed federal judges who presumably would favor religious teachings in public schools. His administration intervened before the Supreme Court in three schoolhouse cases, losing all three (noted above). After five years in office, President Reagan had appointed only one justice to the Supreme Court but had named six judges to the fifteen-member Fifth Circuit appellate bench.[9]

Creationists have anticipated that appointments of friendly judges would contribute to the success of their cause.[10] After the rejection of Louisiana's creationism law, Bill Keith, the Louisiana senator who sponsored the law, expressed the opinion that, "God had delayed the lawsuit this long so that more Reagan appointees could join the bench." Wendell Bird, special defense for the Louisiana Creation Act, also set his hopes on the appellate court. He bypassed the U.S. Supreme Court and appealed for a rehearing by the entire fifteen-member bench. It was not to be. The jurists rejected the review by a bare majority, and the case went to the Supreme Court, where it was defeated on June 19, 1987.

School Prayer Regimen: Religious Indoctrination

Students aren't praying enough—if at all, say the bureaucrats. Other faithful politicians parrot the same words and declare that something must be done. Many parents agree. They want their children to pray but can't find time to help and are seldom able to get organized for prayer in the home.

Most of those urging more prayer are not organizing and pushing for prayers in homes or in churches; they have decided, "Let the schools take care of it." That is the standard, painless procedure. Bureaucrats and busy parents give flippant and easy answers and then expect principals and teachers to organize and conduct a program for school prayer.

The insistent call from some elected officials and other advocates is for "voluntary" prayer in schools. Note the illogic in that call. If prayer depends on governmental edict and is organized and directed by school personnel, it is not voluntary. An important question is this: Who makes the choice? Is it to be the student's volition and choice? Or must our youth, like automatons, conform invariably to the will of public tutors?

Students already have opportunity for voluntary prayer in American schools. Some politicians and other prayer advocates imply that prohibitions exist, so that students are denied free choice and therefore could not utter spontaneous (voluntary) prayer anytime and anywhere. There is no such constraint. Why, then, does America need a statute to legalize prayer in public schools?

There are numerous other important questions that some politicians and parents never take time to consider.

How would prayer sessions be conducted in schools? Would teachers tell children why they should pray? Inevitably, children will receive some answers to that question. And the answers will involve instruction about gods—perhaps a dominant God and various lower personalities—who listen, who desire to hear prayers, who enjoy praise, who answer prayers, who protect the faithful, and so forth.

Prayer is a learned religious exercise. Would instructors tell children how to pray? Specifically what to say? Would they instruct children in prayer-related exercises: bowing, kneeling, or other movements? Would teachers inform children about proper attitudes for prayer and exemplify in their own demeanor those presumably appropriate feelings and emotions, such as reverence, humility, and penitence in respect to specific deities?

All instruction about how to pray would indoctrinate. It would help establish in children's minds ideas about the alleged nature and acts of deity according to specific biases of sectarian opinion. What authorities, such as local pastors, traveling evangelists, or other religious instructors, in addition to school personnel, would be allowed to take part in classroom indoctrination and prayer sessions?

Whose God would be addressed? In our pluralistic society where all religions are constitutionally equal, would indoctrination and proselytizing of Jews, Muslims, Buddhists, Hindus, and other non-Christians be allowed?

In whose name (authority) would prayers be addressed? In the name of a prophet, a saint, some special intercessor, or other intermediary? This is fundamental to religious doctrine and prac- tice. For Christians, prayers are believed effectual through the agency of (in the name of) Jesus Christ, a doctrine certainly offensive to devotees of other religions.

Would American schools tolerate voluntary prayers of the Muslim student bowing toward the holy shrine in Mecca? Or a Tibetan student, of Buddhist parents, spinning a prayer wheel? I submit that, upon hearing of such practices in their local schools, some disturbed parents would transfer their children to "safe" classrooms—or raise a great clamor for changes in school-prayer laws.

State-Sanctioned Religion and "Thicker Skin"

Some Americans believe that citizens of the states are still as free as the colonists before the Constitution to establish whatever religions they see fit. Indeed, a political unit, such as a county or state, could be administered as a parish. The First Amendment's "free exercise" provision, they believe, justifies that claim.

A notorious example of that claim exists in a repealed Alabama law. Chief Judge W. Brevard Hand described specific effects of the free exercise of the Alabama law that established prayer routines in the public schools.[11] He explained that the legislation causes psychological pressure and irritation, as "a necessary consequence," in "non-believers" or in members of a "religious minority." The pressure, he said, flows naturally from state actions, for example, when industrialists come under tough environmental laws, or when unemployed workers (strikers) are subject to state laws that allow strike breakers to cross picket lines. Indeed, everyone at times can feel coerced. This was Judge Hand's remedy:

> The Constitution, however, does not protect people from feeling uncomfortable. A member of a religious minority will have to develop a thicker skin if a state establishment offends him. Tender years are no exception.[12]

While some children pray, others apparently must suffer isolation and embarrassment.

That is not the heritage of freedom promised to American citizens. No child in America should be forced to develop an insensitive mind, deadened by psychological abuse that injudicious officials consider natural and necessary. Fortunately, the Supreme Court rejected Judge Hand's prescription. Alabama's prayer law was unconstitutional.

All branches of government, allied with educational authorities and parents, should work very hard to eliminate such potential tyrannies over young minds. It should be the concern of all citizens that no religious doctrine or practice ever becomes established in the public schools.

Notes

1. B. LaHaye, USA Today, August 17, 1984.

2. J. Falwell, America Can Be Saved (Murphreesboro, Tenn.: Sword of the Lord Publishers, 1979).

3. R. Simonds, National Association of Christian Educators fundraising letter, 1984.

4. P. Weyrich, "Concerned Women of America Monthly Report." November 1988.

5. C. V. Bergstrom and D. Saperstein, "God and Politics," Washington Post, August 26, 1984. Reverend Charles V. Bergstrom is executive director of the Office for Governmental Affairs for the Lutheran Council in the USA. Rabbi David Saperstein is director of the Religious Action Center of Reform Judaism in Washington.

6. Americans should honestly face this question: Should any church, as

a tax-free institution, be allowed to use its property for political activity (such as voter registration drives and electioneering) and continue to maintain its tax-free status?

7. E. Doerr (ed.), "There He Goes Again," *Voice of Reason*, no. 13 (1984): 3.

8. *Aguilar v. Felton*, 473 U.S. 402 (1985); *School District of City of Grand Rapids v. Ball*, 573 U.S. 373 (1985); and *Wallace v. Jaffree*, 472 U.S. 38 (1985).

9. E. L. Larson, *Trial and Error: The American Controversy Over Creation and Evolution* (New York: Oxford University Press, 1989), pp. 172-75, 226 (note 50).

10. Ibid., p. 173.

11. W. B. Hand, *Jaffree v. Board of School Commissioners of Mobile County,* 554 F. Supp. 1104-1130 (1983), p. 1118.

12. Ibid., p. 1118 (footnote 4).

20

The U.S. Constitution and Religion

The First Amendment of the Constitution reflects the clear vision of the Founding Fathers. They charted a worthy and fortunate model for national harmony and freedom. Under constitutional protection, the Church would be safeguarded against the State's interference. And the State would be safe from Church control. It would take strongly organized political apostles of a dominant tradition or a powerful sectarian coalition to establish dictatorial authority that could abolish civil law and destroy the religious and civil liberties of the people.

Separation of Church and State

On September 12, 1984, with the Reagan-Bush election campaign moving into its final weeks, Bill Moyers appeared with Dan Rather on the "CBS Evening News." Apparently mindful of President Reagan's recent appointment of a U.S. ambassador to the Vatican and his intimacies also with Protestant fundamentalists, Moyers noted that believers have rights "to press their view in the public square. And politicians almost always go hunting for votes in the precincts of the faithful." He then asked, "So what's new in this campaign?"

He continued,

This is new. Conservative Catholics and Protestants have openly allied themselves with the Republican Party in a way that threatens to turn the public debate on morality into a partisan crusade and make Mr. Reagan's party the party of religion. That would be a profound change in American politics. . . . In time, nothing but trouble is likely to come of a major party's commitment to the doctrinal triumph of one sectarian notion of God's will for America.[1]

Moyers concluded, "We have in this country an admirable alternative to civil war, and to a holy civil war at that. It's called the Constitution."

Government, Science, and State-Established Religion

The last quarter of this century has witnessed a serious display of religious bigotry and the acceleration of attacks on the U.S. Constitution.

Scientists can take courage that in the text of the Constitution the Founders recognized the importance of "science and the useful arts" for the welfare of the nation. Under Article I, "Powers Vested in Congress. . . . Section 8," are these words:

The Congress shall have power: . . . (8) To promote the progress of science and the useful arts, by securing for limited times to authors and inventors the exclusive right to their respective writings and discoveries.

In contrast to the vested powers of Congress to promote science, no branch of government was empowered to promote religion. Religion was mentioned once to specify that the government's role must remain neutral. And the word "religious" occurred once to emphasize that such tests could not qualify a person for political office. The Constitution did not mention any God, prophet, sacred writing, religion, or any religious institution.

There is no scarcity of quirky notions about the design and nature of the Constitution itself. Walton subscribed to the theocratic notion of a Christian state, "To read the Constitution as the charter for a secular state is to misread history . . . the Constitution was designed to perpetuate a Christian order."[2]

The unfortunate prejudice from the president of the Moral Majority Foundation also reflects the parish mentality:

> So away with the ill-informed, anti-American, anti-Christ activists who tell us that the First Amendment was born of secular seed, designed to ensure a secular America. They have twisted and perverted our precious Christian First Amendment heritage enough.[3]

Evangelist Pat Robertson stated political aspirations for religious-fundamentalist control: "We have enough votes to run the country."[4]

The above quotes represent the mindset of some who believe that evil forces are blocking—only temporarily—the manifest destiny: political dominance by Christian fundamentalists and those who will become their allies.

Attacks on the Supreme Court and Congress

The mid-1980s were ominous watershed years. The Supreme Court justices were being pressured to confirm initiatives that would radically change public school policy. Congress was blamed for failure to counter the Court's rulings against school prayer. And the Court was accused of disdain for religious belief and of fostering atheism in the public schools.

Regarding Supreme Court power, Senator Jesse Helms would thunder, "It's time for Congress to stand up to the Supreme Court [and to] withdraw federal jurisdiction over school prayer."[5]

On August 7, 1985, Education Secretary William Bennett pledged that the Reagan administration would continue pressing for judicial reforms: "The administration in which I serve will continue to press for legislation and, where necessary, judicial reconsideration and constitutional amendment to help correct the [Supreme Court's] current disdain for religious belief."[6]

Evangelist Pat Robertson announced this problem-solving strategy: "If two new justices were appointed, we wouldn't need constitutional amendments regarding abortion and school prayer."[7]

Influential Christian fundamentalists are as convinced in the 1990s as they were in the 1980s that presidential appointments of judges will continue to help advance their cause. James Dobson, head of the organization called Focus on the Family, is an example. His radio show airs fifteen times a week on 1,450 stations in the United States and thirty-five other countries where he reaches millions. He described how Christians plan to win on three fields:

> Our battle . . . will be a battle of ideas. . . . The first is at the ballot box. . . . The second battleground will be fought over the public schools. . . . And thirdly, and please take note of this one: someday when you awaken in the morning and you get your newspaper, the headlines will show that there is a new Supreme Court vacancy. Be ready for that moment.[8]

American history has shown that the wall between Church and State has remained as a bulwark against religious zealots attacking from the outside. One can only wonder how long the wall can endure when elected officials on the inside, in the highest positions of government, are determined—and have deliberately appointed potential accomplices—to help tear down the wall.

The Wall of Separation: Can It Be Preserved?

Most churchmen and churchwomen have recognized that the separation of Church and State is a congenial arrangement. For them, it would be the greatest insult for the Church to profane herself to become the "kept" woman who submits to the support and domination of the State. As for the reverse situation, ruling civil authorities have generally rebelled at the thought of sub-servience to the Church.

With these thoughts about essential separation, let us consider a profoundly disturbing statement that, to some, will suggest an ominous turning point in America's history. The Chief Justice of the Supreme Court declared:

> The "wall of separation between church and state" is a metaphor based on bad history, a metaphor which has proved useless as a guide to judging. It should be frankly and explicitly abandoned.[9]

Justice Rehnquist's written opinion revealed that a crucial part of his dissent rests on two catchwords, "religion" and "irreligion." He mentioned the pair five times, and each time he emphasized that the Establishment Clause does not require the government to stand neutral between religion and irreligion.

We thus have from our Chief Justice a new revelation, essentially a religious dogma, presented as a political moral imperative.

Other officials of the Reagan-Bush administration have taken up the same no-neutrality argument. Attorney General Edwin Meese voiced this speculation: "To have argued . . . that the [First Amendment] demands a strict neutrality between religion and irreligion would have struck the founding generation as bizarre."[10]

Meese's assertion apparently rests on the notion that irreligion is a moral evil to be eradicated and that government is obligated to act in specific ways to eliminate that evil. It is, of course, true that two hundred years ago, as at present, many Americans would have considered the First-Amendment neutrality inconsistent with their religious views. Nonetheless, the wall is there, built to enforce neutrality.

But how shall we answer the above assertions about religion and irreligion? Americans readily understand that responsible government does not take a neutral stand between obedience to law and disobedience. The State must use its investigative, policing, and judicial powers to uphold the law. But it should be obvious that the specific laws upheld are civil and criminal, not canon law.

The government, therefore, has no power to investigate and discipline anyone for irreligion. It cannot correct religious belief or bring censure for disbelief. It cannot compel participation in religious activities or enforce displays of "proper" attitudes. It cannot inflict punishment or require penance for violations of religious codes. Church authorities are quite conscious that such powers belong to them and that disciplinary action can be brought only against church members.

In summary, irreligion, under constraint of the First Amendment, is not an evil punishable under any statuary law; it is a guaranteed option enjoyed by all citizens who exercise freedom of conscience and freedom of choice in all matters of religious faith and practice.

Must Government Establish Religion and Eradicate Irreligion?

Leaders who do not carefully define the words they use may seriously delude themselves and those who follow them. "Religion" and "irreligion" are crucial words that should be clearly understood. To be irreligious is to be "neglectful of religion, lacking religious emotions, doctrines, or practice."[11]

The three words "doctrines," "practice," and "emotions" embrace all possible religious experience and thus help to define irreligion. I suggest several examples with the understanding that they cannot represent precisely any reader's personal beliefs and related experiences.

Doctrines. Several world religions postulate supernatural beings who watch and listen; who experience numerous emotional states (including pity, love, anger, jealousy); who create, curse, punish, and smell the pleasing odor of burnt offerings; and who require blood sacrifices for atonement, hear the prayers of supplicants, but reserve eternal punishment for unrepentant sinners. There are doctrines of inspiration, revelation, angels, devils, saints, chosen people, a universal flood, heaven, and hell.

To be irreligious means that one may disbelieve and reject any or all of the doctrines named above and a host of others not mentioned. If the government cannot be neutral, but must eliminate irreligion, then our officials are obligated to specify doctrines that Americans must believe. Obviously, no official in any branch of government has that authority. Persuading society that religious doctrines represent truth and are therefore binding on everyone must be left to the churches.

Practices. In broad aspect, religious practices range from the solemn and ceremonial (steeped in liturgy and ritual) through the less formal to the wide-open, frenetic orgies of physical and the allegedly direct "spiritual" communion.

Common ceremonial rites among Christians are confirmation, consecration, baptism, and the eucharist or Lord's Supper. Practiced most regularly are prayer, Bible reading, preaching, and—occasionally—fasting, personal testimonials, and exhortation. Other practices may involve liturgical chants, incense burning, candle lighting, manipulation of prayer beads—all used in religious devotions.

We may reasonably conclude that our public servants have no more authority to prescribe religious practices than they have to dictate what people must believe. Regardless of First Amendment prohibitions about advancing religion, however, government officials have repeatedly proposed amendments to the Constitution that would mandate the practice[12] of prayer in public schools.

I pose a question for the future. If classroom prayers were mandated by law, would the Congress and Supreme Court agree that students could be outfitted with prayer beads, devotional guides, candles, incense, religious garb, jewelry, or any other thing from a vast catalog of paraphernalia thought to enhance prayer sessions?

Emotions. The pilgrimage of Pope Paul II, telecast from Poland (NBC, June 8, 1991), gave striking aspects of religious emotion. With profound reverence and submission, the Polish president, Lech Walesa, fell to his knees and kissed the pope's ring. An interviewer asked a Polish woman for her impressions. Her response was, "I'm so moved, I can't think straight." Such is the overriding power of emotion and religious faith.

Sacred persons and ceremonies, sacred artifacts and symbols, holy sites and structures, holy days and celebrations—everything communicated by sight and sound, even touch, taste, and smell—may generate in religious devotees a wide range of emotions. They include wonder, fear, homage, reverence, humility, guilt, grief, penitence, affection, and adoration.

The U.S. Constitution protects lawful religious expression, as well as persons and their property. No intelligent citizen who is sensitive to the civil rights of others would deny them their legitimate faith. Over the whole range of human feeling and expression, however, there is no government formula and no binding prescription for religious emotions that American citizens must experience.

The majority opinion in *Epperson v. Arkansas* provides a comprehensive Supreme Court summary of necessary government neutrality in matters of religion:

> Government in our democracy, state and national, must be neutral in matters of religious theory, doctrine, and practice. It may not be hostile to any religion or to the advocacy of no-religion; and it may not aid, foster, or promote one religion or religious theory against another or even against the militant

opposite. The First Amendment mandates governmental neutrality between religion and religion, and between religion and nonreligion.[13]

The America That Enters a New Century

Citizens would do well to ponder what kind of an America will enter the twenty-first century—if the following activities continue throughout the remaining years of this decade:

• if a dominant political party functions as the party of religion, contributing more and more to a dominating evangelism and to the moral prejudice of sectarian opinion,

• if politicians allied with the conservative Far Right continue pushing for Bible teaching in public schools, prayer in public schools, and public aid to private and parochial schools,

• if presidents misperceive (or disregard) the proper role of organized religion in a secular democracy where, constitutionally, Church and State should remain separate,

• if presidents appoint, and Congress approves, jurists whose purposes and opinions advance sectarian religious goals,

• if presidents, vice presidents, and other officials choose to cultivate the evangelical right-wing community, seeking opportunities before its politically active audiences to endorse their sectarian objectives, to report official efforts to advance religion, and to assure that the government, and they personally, will continue working to establish specific religious beliefs and practices in society,

• if popular but impressionable and unreflective political leaders embrace a mystical and pietistic mix of chosen-nation, save-the-world Messianism combined with superstitious Armageddon, end-of-the-world fantasies,

• if political leaders and allied militant religionists repudiate the "wall of separation between church and state" and persuade the citizenry that government must abolish irreligion by establishing specific religious doctrines and practices throughout society.

The seven conditions stated above are not imaginary scenarios; they have ample precedent in political and religious activity, particularly in the past decade. If powerful religious coalitions that recognize no physical bounds continue to gain power and influence, and find help from eager politicians, then America can expect substantial "theocratic" dominance and abuse of her institutions—and a national turmoil of partisan hatred.

From the history of struggles between religious authority and civil powers, the Founding Fathers understood that our Ship of State must not depart from a straight, undeviating passage. There must be no change in compass bearing that could start the frightening spiral of sectarian bigotry that would sweep America into the vortex of religious hatred and strife.

Notes

1. B. Moyers, "CBS Evening News," September 12, 1984.

2.. R. Walton, "One Nation Under God," in *The Rebirth of America*, ed. N. L. De Moss (Philadelphia, Pa.: Arthur S. De Moss Foundation, 1986), p. 19.

3. J. Falwell, *Moral Majority Report*, September 1985.

4. P. Robertson, *Washington Post*, August 19, 1985.

5. K. Sawyer, "Conservatives Renew Bid for School Prayer Law. Supreme Court Ruling Widely Denounced," *Washington Post*, June 6, 1985, p. A7.

6. W. Bennett, Address to the Supreme Council, Knights of Columbus, Washington, D.C., August 7, 1985.

7. P. Robertson, *Christianity Today*, January 1986.

8. J. Dobson, *Rally for Life*, April 28, 1990.

9. W. Rehnquist, Dissenting opinion, *Wallace* v. *Jaffree* 472 U.S. 38, June 4, 1985.

10. E. Meese, American Bar Association address, July 9, 1985.

11. *Webster's Ninth New Collegiate Dictionary* (Sprinfield, Mass.: Merriam Webster Inc., 1988).

12. Supreme Court Justice Sandra Day O'Connor sided in *Wallace* v. *Jaffree* with the majority who outlawed prayer in Alabama public schools. She understood that the Alabama law improperly crossed the line by endorsing a particular religious practice of prayer.

13. A. Fortas, *Epperson* v. *Arkansas*, 393 U.S. at 103, 104 (1969).

Appendix

Energy Absorption Bands for Atmospheric Water Vapor

Vibration-Rotation Band	Range (Å)	Vibration-Rotation Band	Range (Å)
411	5414-5470	212	6424-6563
203	5665-5769	231	6565-6585
500	5701-5766	103	6845-7128
321	5828-6019	400	6903-7203
401	5830-5999	301	7059-7408
302	5863-5983	221	7099-7480
113	6275-6375	202	7131-7392
311	6408-6626	320	7201-7358

The sixteen light-absorption bands in the visible spectrum are from C. E. Moore, M. J. G. Minnaert, and J. Houtgast, "The Solar Spectrum 2935Å to 8770Å," National Bureau of Standards Monograph 61 (Washington, D.C.: U.S. Government Printing Office, 1966). Over the wavelength range of 5414-7480 Angstroms the total count of atmospheric water vapor lines exceeds 1000. The lines are identified as "Atm H$_2$O" (see Moore et al., 1966, pp. 232-314). In addition, more than 570 lines in that same spectral range are identified only as "Atm." The authors note (p. xix): "Unclassified atmospheric lines are entered as 'Atm,' although they are probably due to the water vapor molecule." The bands are denoted as "vibration-rotation," which identifies mechanisms in the energy absorbing process. Three dominant *vibration* frequencies in the water vapor molecule are active in light absorption. Transitions in *rotational* energy of the molecules are also involved in radiation absorption (K. Y. Kondratyev, *Radiation in the Atmosphere*, International Geophysics Series, vol. 12 [New York: Academic Press, 1969], pp. 108–109).

Index of Names